ELECTROMECHANICAL BUILDING BLOCKS FOR THE MODEL ENGINEER

ELECTROMECHANICAL BUILDING BLOCKS FOR THE MODEL ENGINEER

PAT ADDY

SPECIAL INTEREST MODEL BOOKS

Special Interest Model Books Ltd.
P.O. Box 327
Poole
Dorset
BH15 2RG

© Special Interest Model Books Ltd. 2006

The right of Pat Addy to be identified as the Author of this work has been asserted by him in accordance with the Copyright, Designs and Patents Rights Act of 1988.

All rights reserved. No part of this book may be reproduced in any form by print, photography, microfilm or any other means without written permission from the publisher.

ISBN 1-85486-243-X
EAN 9 781854 862433

www.specialinterestmodelbooks.co.uk

Contents

Chapter 1	Introduction	7
Chapter 2	Basic Electromagnetic Theory	12
Chapter 3	Stepper Motor Drive	16
Chapter 4	DC Motor Drive	34
Chapter 5	The Servo System	67
Chapter 6	Relays	79
Chapter 7	Solenoids & Related Devices	83
Chapter 8	Other Electromagnetic Devices	93
Chapter 9	Interference Suppression	96
Chapter 10	Heatsinks	100
Chapter 11	Fuses & Circuit Breakers	104
Chapter 12	Inputs	107
Chapter 13	Light Emitting Diodes	126
Chapter 14	Speed Measurement In The Workshop	128
Chapter 15	Power Supplies & Regulators	131
Chapter 16	Power Supplies From Batteries	141
Chapter 17	NVRs & Interlocks	147
Chapter 18	Ancillary Test & Driver Modules	155
Chapter 19	Basic Electronic Building Blocks	160
Chapter 20	Practical & Cost Effective Building	166
Chapter 21	Etching Processes	175
Chapter 22	Using Stripboard For Prototypes	180
Chapter 23	Pin Outs & Specifications	182
Chapter 24	Information Sources	187

CHAPTER ONE

Introduction

The last twenty years have seen a great leap forward in the world of electronics. Inexpensive dedicated microprocessors and the power MOSFET have had a great effect. The swift progress of the PC has put large amounts of fast processing power at the hands of the designer. Over the last decade or so we have seen the rise of two new 'sciences' namely mechatronics and animatronics. Both these areas are concerned with the control of movement by electronics. Although the stepper motor and the DC motor have changed little in their design, new materials and new manufacturing methods have brought costs down to the extent that stepper motors have replaced small AC motors in many applications. The two events that have probably moved DC motors forward the most are the growth of the cordless power tool in the DIY market and the use of small powerful DC motors in the flying and other model fields.

These types of motors are usually low voltage and are therefore inherently safe from an electrical point of view. Safety is still paramount from a mechanical point of view when motors weighing a matter of ounces can take voltages and currents equivalent to an input of five horsepower from a battery pack, albeit for a very short duration.

The object of this book is to provide the person with an interest and possibly knowledge of mechanics a number of building blocks for use with electromechanical devices.

This covers the driving of stepper motors, DC motors and a wide range of other electromechanical devices in the form of blocks that are practical, work well and are easy to build. The blocks can be mixed and matched to cover almost any application that is likely to arise using these devices.

Making a motor or other electromagnetic device operate is only half the story. It is necessary to have interfacing with other devices.

The book goes on to cover the control of these blocks and how physical data such as position and movement can be translated into terms that can be used for control. It then moves further in this direction to look at how other terms that affect these electromechanical items such as temperature can be monitored and used in control situations.

Aspects such as safety and the prevention of electromagnetic radiation and interference are also covered.

There is a small amount of theory and mathematics but this is only taken to the level of 'need to know' to be able to use the practical concepts within the book.

The book then goes on to show a number of practical examples of using the blocks. On top of this, theoretical designs are presented in block form that will hopefully encourage people to experiment, develop and use the ideas in there own designs. The circuit designs shown as block diagrams can all be built using the fully described and component detailed 'building blocks'.

All complex movement can be broken down into a series of simple steps.

The book then moves on to explore aspects of practical building such as the approach to breadboard and PCB making.

Topics such as etching discs to make speed and position sensors are also covered.

The words drive and control in electronics are often used synonymously. Within the context of this book it is the intention to attempt to use the word drive or driver to be the hardware that switches power to make a unit work. Control is being used as any device that makes the driver operate.

Safety

Safety is of paramount importance when undertaking any form of engineering activity. Motors even quite small ones particularly when geared can supply sufficient power to easily 'chew' any unsuspecting finger.

Any rotating object has inertia that can cause damage if it breaks free.

Most low power circuits unless totally battery driven will have a mains AC supply somewhere. Electronic components although running within specification can be hot enough to cause burns. Low impedance batteries and charged capacitors can cause a serious burn if a watchstrap or jewellery bridges the terminals.

Even quite small electronic devices can disintegrate in dramatic style if their specifications are exceeded or wired incorrectly. I have seen power transistors that because of a short circuit condition overheat and explode throwing hot pieces of the body encapsulation many feet at high speed.

There are also secondary effects that are caused by the surprise of another event. And often these events later may seem funny but at the time could have had very serious consequences. Many years ago where I was working, a colleague turned up one morning with two large lumps on his head. He had been adjusting a CRT driver unit in a confined place. Unfortunately he was distracted and his hand brushed the EHT circuit. The resulting shock caused him to jump upwards, unfortunately three inches above his head was the heavy steel frame of the machine. The obvious happened and head and steel plate met at speed. Unfortunately this is where things took an even worse turn. On his way back down from the collision he put his hand back onto the EHT circuit with the inevitable further collision between head and steel plate. Fortunately experience now kicked in and he avoided the EHT circuit the third time.

The main rules of safety are relatively simple Assess the situation and any potential dangers fully.

Remove anything in the vicinity that could have a secondary effect.

Everything in life carries some risk but do not take this beyond what is reasonable in the circumstances.

If using any chemicals read, understand and abide by any relevant safety information.

Testing units

When you build a mechanical unit it is possible most times to assess it whilst it is stationary and then 'turn it over' slowly by hand. Any problems found can usually be sorted before running full power tests.

Electrical and electronic circuits are different in that you cannot make the circuit work slowly by hand.

There are a number of tests that can be done and these will mainly involve the use of a meter. Short circuits are a major problem that have a catastrophic effect but these can usually be traced fairly easily. Many of the circuits in the book can be set up, tested and used as individual blocks. There are also a number of inexpensive and simple to build test modules included. These are worth making if it is the intention to build a number of the projects.

I have been involved with many repair workshops and research projects over the years. It quickly became apparent that no matter how much static testing is done with some designs it is only when the power is connected that you know if it works.

None of the projects in the book should come into this category.

The technique used and probably still in use in many workshops was based on the long stick philosophy. This applied mainly to mains driven equipment and involved plugging the unit into the mains and whilst hiding behind the bench, switching the socket on with a long stick. The absence of a loud bang usually meant success. I still have my own version of the long stick but now upgraded. It consists of a box with a socket, changeable MCBs, a RCD and a relay. A long lead switches the unit on via the relay.

I hope this just helps to illustrate the point that it is much better to be safe than sorry.

Safety can also be a matter of common sense within designs. See also the section on NVRs and interlocks.

Static precautions

Many components particularly ICs are susceptible to static charge. Walking across a carpeted or synthetic floored room can generate many thousands of volts of static electricity. It takes only a small amount of charge above the normal working voltage to destroy or degrade a component.

During the late 1970s static precautions started to be taken seriously outside of the actual manufacturing environment by major companies. It was rumoured that just by issuing the repair personnel basic wrist straps and by using antistatic packaging one multinational company cut its repair bill over a two-year period by 35%. Static can be 'cured' on two levels. The first method is by not allowing it to be generated and dissipating any static produced is the second method.

There are three kinds of work surfaces, non-static generating, conductive and dissipative. Natural materials such as wood do not produce static although if varnished, polished or painted the properties may be negated. Conductive surfaces are normally of a high resistance material and the resistance can be measured across points on the surface. Because of the measurable leakage resistance it is not possible to power test units on a conductive surface because of the possibility of short circuits. Dissipative surfaces conduct between the upper and lower layers not across the upper layer. Dissipative surfaces can be used for live testing. The disadvantage is the cost that can be many times the other types of surface.

Commercial electronic premises may have dissipative surfaces and conductive floors and personnel with cotton clothing and conductive ankle straps. The air and humidity may also be controlled to prevent any possible static build up.

For the home constructor basic equipment is usually sufficient. This consists of a wrist strap and a connection via a 1 MΩ resistor to an earth point. The resistor is to prevent risk of shock if the lead accidentally contacts high voltage. Conductive mats can be bought but a wooden surface is usually as good. Basic rules need to be followed. Do not open any static sensitive components except at the static free area they will be used at. Do not walk about with assembled PCBs unless they are protected with appropriate static free packaging. Soldering irons need to be of a 'safe' type i.e. they are designed for this type of work and do not have high voltage leakage from the tip. Always use a wrist strap when handling static sensitive components. Problems caused by static do not always show by immediate failure. If degraded they may work for a long period of time but premature failure is almost certain and the effect of a degraded component can have catastrophic effects on the rest of the circuit.

Accuracy, resolution and repeatability

This is an area that causes problems. Accuracy, resolution and repeatability although fundamentally related are totally different measurements. Misleading information can be easily conveyed by the inappropriate use of terminology. This may be inappropriate marketing or an effort to imply that a piece of equipment behaves better than it actually capable of.

Consider a device such as an electronic compass reading from 0° to 360° in 1° steps.

A phrase that could be used is 'accurate to 1 degree'. This may be correct in everyday language but in scientific or engineering terms is totally incorrect. In reality it has a resolution of 1° i.e. it can measure in steps of 1° but this does not necessarily mean that the reading is accurate. Accuracy in this instance is the relationship of the compass reading to the true reading i.e. if the compass indicates 15° but the true reading is 17° this is an inaccuracy of 2°. Accuracy is normally specified as a +/- error e.g. +/- 2°.

Repeatability is the ability to return to the same position or return the same reading on subsequent operations.

The same compass with an accuracy of +/- 2° and a resolution of 1°, if used to take a reading that was 15° could return a reading of 13°, 14°, 15°, 16° or 17°.

If the compass always returned a reading of 17° for a reading of 15° then the repeatability would good even though there was still a 2° inaccuracy. These concepts apply to most items and can put practical limits on a design whether they are mechanical or electrical/ electronic. As with the compass there would be little point designing a display that would give a read out with a resolution to 0.01° based on an accuracy of +/- 2°.

A mechanical example is the microstepping of stepper motors. This can have practical advantages in the smooth running of a stepper motor but the number of micro steps compared to the accuracy of each full step and the load on the motor can mean that true positional accuracy and repeatability of positioning does not match the expectations implied by the electronics.

Tolerances

All components are manufactured with tolerances whether they are mechanical or electrical/ electronic.

When designing or using any circuit these must be taken into account. The tolerances on most circuit components are such that they should have little effect on most practical designs.

The times when problems are noticed are likely to be with potentiometers in compare circuits. Servo or feedback circuits can have problems due to component tolerance if they cause a restriction in the voltage range such that the transition points cannot be crossed and hence switching does not occur.

Data sheets and technical specifications

Manufacturers data sheets and technical specifications are probably the best source of information once a possible suitable component is found. Most of this information is available via the web from the manufacturers website. Unfortunately the level of information varies. Some of the data sheets are written in technical jargon with pages of graphs and very little practical information. Other manufacturers provide pages of practical circuits and applications often with component values. Many components have equivalents made by other manufacturers and it is often worthwhile searching for these other manufacturers and visiting a number of websites. A point worth bearing in mind is that data from even the most reputable of manufacturers can contain errors. When using data sheets, it is wise to read all the data thoroughly and gain sufficient knowledge of the product so that circuits presented can be verified. It is not unknown for circuits to contain wire joins when these should be crossing points. This may prevent correct working of a circuit or in the worst case when using low impedance power supplies may cause destruction of the product or associated components.

Other sources of information are supplier catalogues. Some of the large manufacturers who a few years ago were unwilling to supply their large and probably costly to produce catalogues to anyone who they did not think would spend a lot of money with them are now willing to give away catalogues on CDROM. These often have links to a database area about the components they sell. Once a component has been found it is often worthwhile putting the part number into the web. This usually sources numerous sites where people are offering 'help' about how they used that part. Sometimes novel ideas may come as a result.

The one point that should be borne in mind with any technical information is the credibility of the source. Even manufacturers data sheets sometimes contain minor errors about their own products sometimes as a result of transferring information to print. Other sources may contain information that is totally inaccurate or at the worst potentially dangerous. Even if a circuit is from an apparently credible source it is worthwhile obtaining the data sheet to verify the information to your own satisfaction.

CHAPTER TWO

Basic Electromagnetic Theory

All the devices in this book can be understood with the minimum of physics knowledge. The facts are discussed in this chapter only to a level that is necessary to make the units work.

Conventional current flow

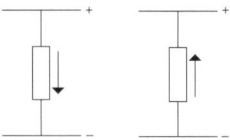

Fig.1 shows conventional current flow. This is regarded as flowing from positive to negative hence circuit diagrams are shown with voltage lines in ascending magnitude with the current flowing downwards. In reality current flow is the movement of electrons and because electrons are negatively charged they flow from negative to positive. This makes no difference in practical terms.

Current flow

There are a number of basic terms used and this can be explained with an analogy of water flowing through a pipe.
Voltage is the 'pressure' applied. This forces the current flow. The voltage is measured in V(Volts).
Current can be likened to the amount of water that flows in the pipe. It is the current flowing that makes the work happen. The current is measured in A(Amps).
Resistance is has the name suggests the part that resists flow of current. With the water analogy the larger the bore of the pipe the easier and greater will be the flow of water for a given pressure. To move the same amount of water through a smaller pipe we need to increase the pressure or in the case of electricity to increase the voltage. The Resistance (R) is measured in ohms (Ω).
There is a direct relationship between the pressure, the bore of the pipe and the flow. This relationship also applies to electricity relative to voltage, current flow and resistance.
Ohms law states that the ratio of the potential difference between the ends of conductor to the current flowing through it is a constant at a fixed temperature.
Potential difference is voltage difference and the reference to a conductor also applies to a resistor.
This is easily explained by the following diagrams

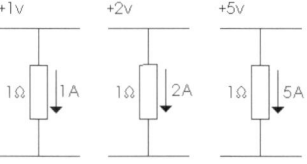

Fig.2 shows current flowing through a resistance of 1ohm. When the voltage is increased the current increase proportionately.
This is given by the formula
Volts equals Current multiplied by Resistance or by the formula V=IR.
Applying this to the diagram
1 = 1 x 1 and 2 = 1 x 2 and 5 = 1 x 5

Anti-static wrist straps with lead and connector for a 13amp socket.

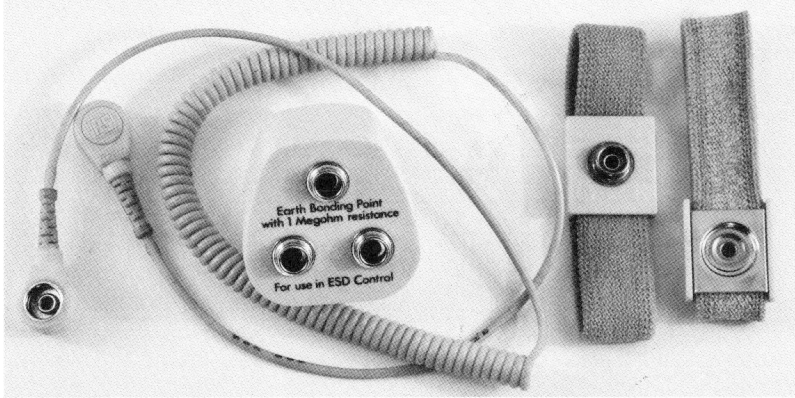

Fig.3 shows the VIR triangle. This is a simple

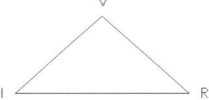

way of manipulating the formula for different unknowns.
If you place a finger over the V then I and R are on the same line and hence multiplied. If you place a finger over the I then V is over R and hence V is divided by R, repeating this for the three possibilities we get
V = IR, I = V/R and R = V/I

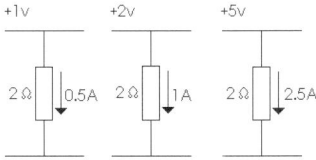

Fig.4 shows the result if the resistance is increased to 2 ohms.
1 = 2 x 0.5 and 2 = 2 x 1 and 5 = 2 x 2.5

Power

Power is the amount of work done or consumed by the device. Power is measured in Watts. The relationship between power, voltage and current is expressed as power equals current multiplied by voltage or by the formula P = IV. This has a major effect when choosing components.
Fig.5 shows voltage flowing through a resistor as in the relationship between voltage, current

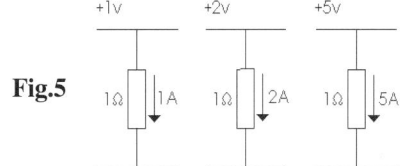

and resistance.
Applying the formula to each case
1 = 1 x 1 and 4 = 2 x 2 and 25 = 5 x 5
This shows that for an increase in voltage the power increases by the square of that voltage for a fixed resistance e.g. doubling the voltage gives four times the power.

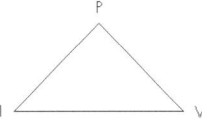

Fig.6 shows the PIV triangle. This is a simple way of manipulating the formula for different unknowns.
If you place a finger over the P then I and V are on the same line and hence multiplied. If you place a finger over the I then P is over V and hence P is divided by V, repeating this for the three possibilities we get
P = IV, I = P/V and V = P/I
When looking at a motor there are two power considerations. The first is the power consumed

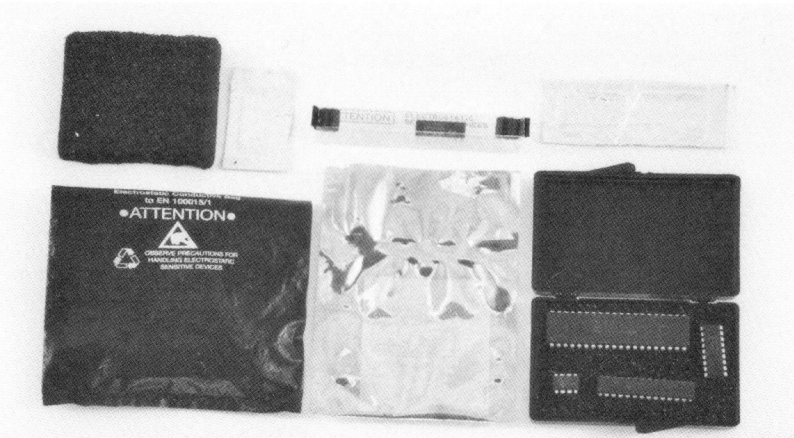

A range of anti-static packaging.

and the second is the power output. The power output is always lower than the power input. The percentage ratio is the efficiency of the motor. These are normally measured at a range of loads and other criteria e.g. temperature.

Resistance and reactance

In the earlier explanations it was assumed that resistance was a finite amount. There are two effects that will concern us when dealing with wound components.

The first is the heating effect. When current flows through a conductor there is a rise in temperature caused by that flow and also an increase in resistance with temperature. This is why motors and solenoids become warm and produce less power the warmer they become. The second effect that occurs in a wound component is reactance. The coil has a resistance but it also has reactance that is the effect of resisting change in current flow. When a voltage is put across a wound component the current rises but the coil attempts to resist the rise. The lower the voltage and the higher the magnetic effect of the coil the longer will be the rise time of the current. This effect will occur inversely when the voltage is removed. This may not seem to be too great a problem but it is probably the major limitation on the speed of a stepping motor. Wound components are classed as inductors and inductance is measured in units of Henrys.

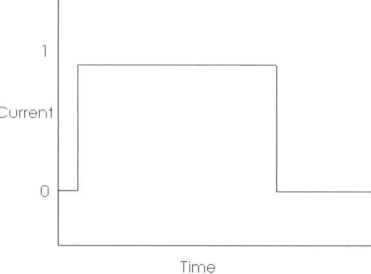

Fig.7 shows the rise and fall of current in a pure resistance.

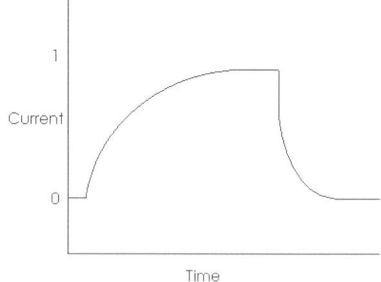

Fig.8 shows the rise of current in an inductor. The actual curve will depend on the inductance of the circuit and the applied voltage.

Electro-magnetism

When current flows along a conductor a magnetic field is produced. Winding the wire around a former concentrates the field and winding the coil around certain metals concentrates the field further. This principle applies to all electromagnetic devices.

A further principle that applies to electromagnetic devices using permanent magnets or electromagnets in their workings e.g. motors is Fleming's left hand rule.

Fleming's left hand rule states that if an electric current is passed through a conductor that is positioned perpendicular to the to the direction of a magnetic field, then that conductor will experience a mechanical force tending to cause it to move in a direction that is perpendicular to both the current flowing through the conductor and the direction of the magnetic field.

It is called the left hand rule because if you hold the thumb, first finger and second finger of the left hand all perpendicular to each other, the relationship becomes first finger is field, second finger is current and the thumb is motion.

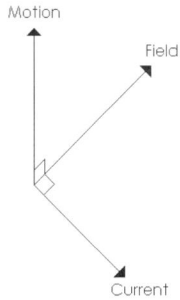

Fig.9 illustrates Fleming's left hand rule

Circuit diagrams and schematics

This can be a confusing area because of the various and often mixed conventions. Even phrases have become mixed. Circuit diagrams tended to be English influence and schematics were of USA influence. Logic symbols were drawn differently but the tendency nowadays is to use a limited range of boxes with the logic type written on. This is probably due to the PC and the international component and machine market. The one point I always clarify before attempting to understand a diagram is the wire joining and crossing convention.

Fig.10 shows various types and mixes of crossing and wire joining conventions.

Switching blocks

The electronic switches used may be transistors, Darlington transistors or MOSFET types. Except where a specific type is discussed they are shown in the form of blocks.

In electronics there are basically two forms of the above devices. The first is 'P' type, this conducts when a negative signal is applied to the control input. The second is 'N' type, this conducts when a positive signal is applied to the control input.

The input level varies with the device and also the permitted current flow to the control input.

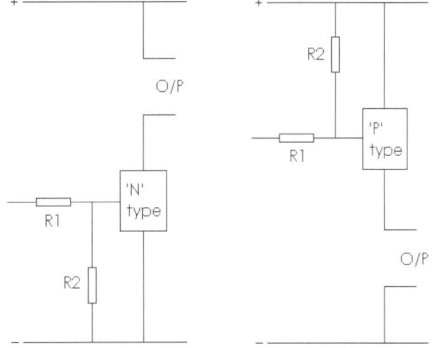

Fig.11 shows the conventional way of connecting 'N' and 'P' type devices. R1 prevents excess current flow to the control input and R2 holds the control input in an off state. Some devices require so little current for switching purposes that the input control can 'float' because of residual charge. R2 ensures that when the input voltage is removed any residual charge is dissipated and the device switches off cleanly. Induced spikes can also cause erratic switching. This may need screening, cable re-routing or bypass capacitors. This is explained in the suppression section.

Basic Electromagnetic Theory 15

CHAPTER THREE

Stepper and Servo Motor Drive

The stepper motor

Stepper motors have been about for many years and the technology has changed little. The stepper motor is basically a number of magnets or metal segments surrounded by a number of coils. Magnetic fields caused by changing the current or current direction in a series of steps through the coils causes the rotor to follow the movement. There are a number of features of stepper motors that need to be understood if they are to be used successfully.

Types of stepper motor

There are many shapes and designs of stepper motor in the engineering world.

Variable reluctance

This type of motor uses an iron non-magnetised core. The core sections are attracted to the stator coils. The number of coil windings can vary and hence the step angle but step angles tend to be relatively large typically in the order of 30°. Variable reluctance motors can usually be distinguished by a number of coils attached to a common point but using this criteria the four-coil type can be mistaken for a unipolar type with a single common point. This does not matter because the method of driving is the same as the unipolar wave drive. The number of coils determines the step sequence. The other possible giveaway with the variable reluctance motor is when the rotor is turned with no power on it turns freely with no 'cogging' or only that caused by residual magnetism.

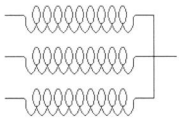

Fig.12 shows the winding of a typical variable reluctance motor.

Permanent magnet

These motors can be unipolar or bipolar and have two coils. Or two sets of coils in the case of a biflar wound motor. When the rotor is turned with no power on it turns with a distinct 'cogging' effect caused by the magnetic poles of the rotor being attracted to the stator poles. The number of magnetic poles it is possible to 'fit' on the rotor limits the step angles. The step angle may typically be 7.5°.

Hybrid

The hybrid stepper motor as the name implies combines the technology of the variable reluctance and permanent magnet stepper motor. The construction is similar to the permanent magnet motor but shaped metal segments on the stator and/ or the rotor are used. These effectively multiply the number of magnetic poles. When the rotor is turned with no power on a 'cogging' effect may be felt caused by the magnetic poles of the rotor being attracted to the stator poles but may feel less than the cogging effect of the equivalent permanent magnet motor. The step angle may typically be 1.8°.

5 Phase 6 wire stepper motors
These have five coils with a common connection at one end of each coil. They are unipolar motors and are often of the variable reluctance type. Usually only one winding is on at any time. The table shows the coil switching of each coil or connection for each step.

5 phase 6 wire stepping
See tables overleaf.

3 phase 4 wire stepper motors
These have three coils with a common connection at one end of each coil. They are unipolar motors and are often of the variable reluctance type. Usually only one winding is on at any time.
The table on Page 20 shows the coil switching of each coil or connection for each step.

5 phase 5 wire permanent magnet stepper motors
These are less common types of motor. The windings have no common point but are wound in a cyclic fashion with a connection being brought out from each coil junction. Coils are switched in sequence by an electronic controller. They can produce high torque from a small physical size because all but one of the windings is switched on at all times. The step angle may be as small as 0.72° equivalent to 500 steps per revolution.
The table shows the polarity of each terminal or connection for each step.

3 phase 3 wire permanent magnet stepper motors
These are essentially the same as the 5 Phase permanent magnet stepper motors but with 3 coils. These stepper motors may not be used in true stepping mode but in continuous run. The main use recently for this type of motor is for large flying models. The motors are used as out-runners where the rotor is the outside case. This has a number of advantages such as producing higher torque because of the rotor size and they can be run at high speed because the rotor magnets are not likely to break away under centrifugal force because they are constrained by the rotor. A motor that is the size of your fist can produce an output as high as 5 HP.

The table shows the polarity of each terminal or connection for each step.

The unipolar and the bipolar stepper motor
The unipolar motor as the name suggests relies on the current moving in one direction through the coils. These usually have paired windings with one centre tap i.e. six wires or a common connection for all four coils i.e. five wires. Most unipolar motors are in fact biflar wound with wires joined to give a single centre tap or two separate centre taps. Biflar winding means winding two wires simultaneously to give two separate coils.

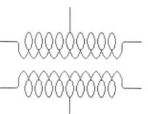

Fig.13 shows the winding of a unipolar motor.

Fig.14 shows the winding of a bipolar motor.

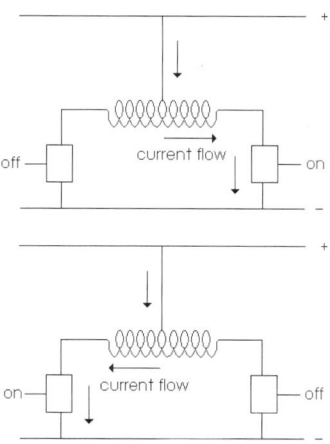

Fig.15 shows the winding for a biflar wound motor with centre tap configuration.
The motor with four separate coils i.e. eight wires is most likely biflar wound. This is useful because the motor can be configured as unipolar

A range of small stepper motors.

or bipolar and the coils wired in series in unipolar mode or series or parallel for bipolar mode. With unipolar motors a set sequence of activation of the coils provides movement in either direction. Switching the current is simple because the polarity is constant.

The bipolar motor is more difficult to drive because it has only two coils and the step sequence also changes the direction of the current flow. A unipolar motor can be run as a bipolar motor by ignoring the centre tap provided the coils do not have a common centre tap i.e. a six-wire motor. There is usually a gain in torque because the complete winding is used. The drawback is that for example a unipolar motor with 5V and 1A coils used as a bipolar winding will require a higher voltage to achieve the same step characteristics. The coils act in the same way as resistors for series and parallel connection. By Ohms law doubling the resistance will require twice the voltage to achieve the same current. Because of the greater field strength it may not be necessary to double the voltage. Bipolar drive is usually achieved by using 'H' bridge configuration. With both types the most common are steps of 1.8° i.e. 200 steps per revolution and 7.5° i.e. 48 steps per revolution. With both types half-stepping can be used which double the number of steps per revolution. Micro stepping is also possible with some types of motor and driver combinations. This allows a larger number of steps per revolution but the resolution and repeatability can vary with the type of motor and drive. In full step mode it is possible to provide current through one coil or through two coils simultaneously. The advantage of two coils is an increase in torque of about 50% but the motor will use twice the power. The effect on the motor is also the relative position of the output shaft; this is effectively offset by a half step between the one coil activation and the two-coil activation.

The choice of one or two coil activation is not possible in half step mode because half step is achieved by alternating between one coil and two-coil activation. The average power consumed will be 150% of single coil full step activation but the torque will depend on whether a single or two-coil step is happening.

Using a stepper motor in bipolar 'H' bridge mode requires a greater component count than with unipolar switching. If MOSFETs are used for switching high power motors the unipolar drive will require a minimum of four MOSFETs and the bipolar drive will require a minimum of eight MOSFETs.

There are a number of control ICs and driver ICs available for use with a stepper motor. These range from the simple SAA1027 (now obsolete) through a number of higher featured controllers e.g. the L297. The drivers available are usually multi purpose devices in that they can be used for other drive applications such as relays and solenoids e.g. L293D, L6203 and L298.

The stepper motor is an inductive device and therefore there is a finite time for the current to rise to its maximum, the larger the winding the greater the inductance. This sets the speed limit for the motor if missed steps or slip are to be avoided. The usual method to make the current rise faster is to increase the voltage applied to the windings. This will also increase the maximum current and can damage the windings. A resistor is placed in the circuit to set the maximum current. The resistor used was normally a wire wound type and these can sometimes be obtained with a non-inductive winding. This limits the maximum current but causes only a minimum effect on the current rise time.

The second method is to use a chopper circuit and these are built into a number of ICs including the L297. These use a relatively high voltage; some drivers use up to 50+ volts for a coil rating of 5V. The current rise time is very quick and the motor is prevented from burning out by the chopper circuit. A resistor is used to indicate to the controller the maximum current and the chopper circuit will switch off each time the current reaches the maximum. This type of chopper circuit can be used in micro-stepping circuits. These are more complex but use the same type of technology to balance currents in sets of coils e.g. if one coil has twice as much current flowing as the coil next to it the rotor magnet will be more attracted to one than the other and will take up a position relative to the field strengths. This is in effect using the digital stepper motor in an analogue mode. The theory is very good but the resolution and repeatability are very much dependent on the current regulation and the quality of the stepper motor. The main advantage is the loss of jerkiness of the drive.

The main disadvantage of the chopper type drives is the possibility of interference that can be put onto the logic and power lines and very careful design is necessary especially when using sensitive high speed controllers in a system because the chopping can be picked up as changes in logic levels.

Another technique that can be used but is not widely seen is voltage level shifting. This is probably not used because of the availability of inexpensive dedicated control and driver ICs. These usually control up to a maximum of about 2A. The voltage shifting technique is a useful technique for stepper motors requiring larger coil current than is available from these dedicated ICs. The microprocessor is a device that can be used for the total control of stepper motors and can extend the use of voltage level control from two levels to multi level. Extra facilities such as temperature monitoring of the motor can be built in to prevent any risk of overheating and burn out. Temperature is also important with permanent magnet stepper motors because if a hot motor is turned against the direction of drive when the coils have current flowing, demagnetisation can occur.

If a large stepper motor is to be used at a high stepping rate there is a problem of the inertia of the motor both when starting and stopping. This can result in missed steps, slip and overstepping. It is necessary to ramp up the speed to maximum and similarly when slowing down. The microprocessor can be used to extend the simple two level control to a number of levels. The next

5 phase 6 wire stepping

Step	Coil 1	Coil 2	Coil 3	Coil 4	Coil 5
1	ON				
2		ON			
3			ON		
4				ON	
5					ON
6	ON				
7		ON			
8			ON		
9				ON	
10					ON

3 phase 4 wire stepping

Step	Coil 1	Coil 2	Coil 3
1	ON		
2		ON	
3			ON
4	ON		
5		ON	
6			ON

5 phase 5 wire stepping

Step	Terminal 1	Terminal 2	Terminal 3	Terminal 4	Terminal 5
1	+	-	+	-	+
2	+	-	+	-	-
3	+	-	+	+	-
4	+	-	-	+	-
5	+	+	-	+	-
6	-	+	-	+	-
7	-	+	-	+	+
8	-	+	-	-	+
9	-	+	+	-	+
10	+	-	+	-	+
11	+	-	+	-	+
12	+	-	+	-	-
13	+	-	+	+	-
14	+	-	-	+	-
15	-	+	-	+	-

3 phase 3 wire stepping

Step	Terminal 1	Terminal 2	Terminal 3
1	+	+	-
2	+	-	-
3	+	-	+
4	-	-	+
5	-	+	+
6	-	+	-
7	+	+	-
8	+	-	-
9	+	-	+
10	-	-	+

simple progression is use a high start voltage then switch to a lower voltage for stepping when the motor is up to its current rating then finally to a low 'hold' voltage.

There have been a large number of dedicated driver 'chips' both for the control and drive of stepper motors. Most of these driver 'chips' are designed for motors of about one to two amps current. Unfortunately most electronic components have a finite manufacturing life. This may not be a problem with components such as transistors and MOSFETs where a substitute will be available.

In these days of mass manufacturing integrated circuits may be designed for one specific project e.g. a disk drive stepper motor driver. When the project reaches the end of its manufacturing life the individual components may also cease manufacture. This causes problems for maintenance companies and for anyone who may have used the individual components in one of their own projects. There are a number of companies around the world who make a good living sourcing and supplying obsolete components. Unfortunately they are rarely interested in the individual who only wants one or two components. The price also escalates with the scarcity.

This is the case with a number of the earlier stepper motor 'chips'. In the case of the disk drive cited as an example, this unit only needs to be 'pin compatible' with the PC. A manufacturer can change design at any stage of production using totally different components.

 This type of unit is generally regarded as 'throw away'. This can be disastrous for the homebuilder or low volume producer who may have used one of the components in a design. This becomes a problem when a repair or another unit is required. The only outcome may be the scrapping of the existing unit and a rebuild using a new design.

Over the last few years in my electronic design work, I have tended to move away from the highly specialised 'chip' if there was any other option. Much of my work is now based around programmable microprocessors with discrete components such as MOSFETs for drivers. Most of the actual integrated circuits I use are 'type' components e.g. OP amps and basic logic gates. The actual component may become obsolete but a compatible or similar type will be available causing the minimum of circuit changes. In the case of the programmed microprocessor these can be designed to be mostly 'plug and play' with existing circuitry or in the worst case be built on to a plug in carrier PCB.

Tables on the facing page illustrate:

Unipolar full step one coil (wave) drive
Unipolar full step two coil drive
Unipolar half step drive

The drive for bipolar motors takes the same format as for unipolar but there is no common point on the coils. It is therefore necessary to reverse polarity of the coils not just switch the coils on. The tables show the polarity of each coil for each step.

Bipolar full step one coil (wave) drive
Bipolar full step two coil drive
Bipolar half step drive

The output from the controller to the driver is the same in unipolar and bipolar mode. The difference is straight switching in the case of unipolar motors and the 'H' bridge switching for bipolar motors.

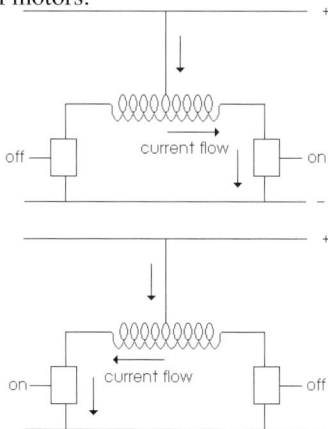

Fig.16 shows the current flow in one section of a unipolar driver.

Unipolar full step one coil (wave) drive				
Step	ØA	ØB	ØC	ØD
1	ON			
2			ON	
3		ON		
4				ON
1	ON			
2			ON	
3		ON		
4				ON

Unipolar full step two coil drive				
Step	ØA	ØB	ØC	ØD
1	ON		ON	
2		ON	ON	
3		ON		ON
4	ON			ON
1	ON		ON	
2		ON	ON	
3		ON		ON
4	ON			ON

Unipolar half step drive				
Step	ØA	ØB	ØC	ØD
1	ON			
2	ON			ON
3				ON
4		ON		ON
5		ON		
6		ON	ON	
7			ON	
8	ON		ON	

Bipolar full step one coil (wave) drive				
Step	Coil 1a	Coil 1b	Coil 2a	Coil 2b
1	+	−	−	−
2	−	−	+	−
3	−	+	−	−
4	−	−	−	+
1	+	−	−	−
2	−	−	+	−
3	−	+	−	−
4	−	−	−	+

Bipolar full step two coil drive				
Step	Coil 1a	Coil 1b	Coil 2a	Coil 2b
1	+	−	−	+
2	+	−	+	−
3	−	+	+	−
4	−	+	−	+
1	+	−	−	+
2	+	−	+	−
3	−	+	+	−
4	−	+	−	+

Bipolar half step drive				
Step	Coil 1a	Coil 1b	Coil 2a	Coil 2b
1	+	−	−	−
2	+	−	−	+
3	−	−	−	+
4	−	+	−	+
5	−	+	−	−
6	−	+	+	−
7	−	−	+	−
8	+	−	+	−

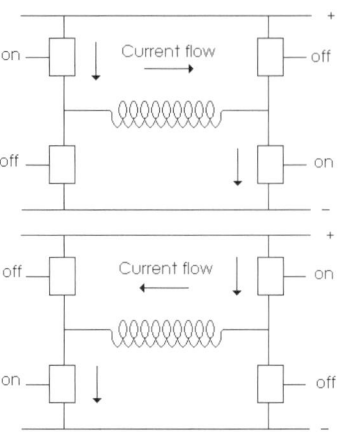

Fig.17 shows the current flow in one section of a bipolar 'H' bridge driver.

Voltage spike diodes

This is one of the commonly misunderstood areas of practical inductive device drive. Motor windings, relays, solenoids and similar devices are all inductive. When a voltage is switched across these devices the inductance resists the change in current. The greater the winding for a given core material the greater the inductance and hence the greater the magnetic field but the greater the resistance to the current change. With DC voltages that are being used with the circuits discussed the current will rise until it reaches a steady state that is dependent on the resistance of the circuit.

When the voltage is turned off the collapsing magnetic field produces a voltage spike opposite to the originally applied voltage. The duration and size of the spike depends on a number of factors such as resistance of the circuit and the applied voltage. This spike can be many times the value of the applied voltage and can cause damage or destroy the switching devices. The most common way of getting rid of this spike is to use an inverse connected diode across the winding.

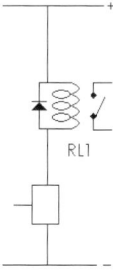

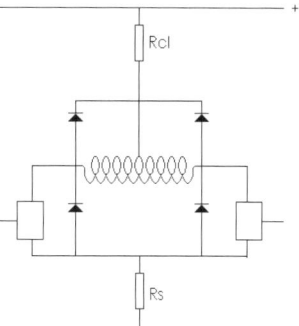

Fig.18 shows a simple single coil such as a relay or solenoid.

The diode prevents current flowing in normal operation but provides a path for the opposite polarity spike. The type of diode chosen is critical. It must be capable of handling the magnitude of the voltage spike and current equivalent to the motor winding or coil current. The diode must also have a switching speed that will allow the collapsing field to dissipate before the voltage reaches a magnitude that can damage the switching components. I have seen a number of circuits using diodes such as the 1N400X series; these are relatively slow devices and are probably all right for some slow current fall circuits of low inductance but they are not suitable for these motor applications. The effect can be that the circuit fails when certain types of driver switch are used but works successfully with other types. The circuit may fail if power transistors are used but work with some type of power FETs. Some power FETs break down non-destructively at a high voltage and can therefore absorb the spike without failure. The usual indication is a rise in temperature of the device. Other methods can be used instead of diodes e.g. capacitors but these other methods are often only practical with specific motor types and specific operating parameters therefore the diode method is being used here as a 'universal' method.

Fig.19 shows a diode set up for one coil pair of a unipolar motor. Rs is the current sense resistor and Rcl is the current limiting resistor

The extra diodes connected to ground via Rs are required because the centre tap of the winding is at a fixed voltage this causes the non driven half of the coil to act like an autotransformer and can induce an inverted voltage across the unused driver. This can be fatal for some types of driver switch.

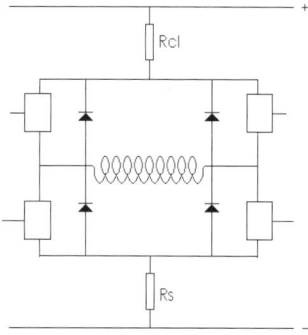

Fig.20 shows a diode set up for one coil of a bipolar motor

The extra diodes are required in this case because the reversal of voltage and current flow through the 'H' bridge circuit.

Problems occur with this diagram in two areas. The first is that the ground of the switches is not the ground of the current sense resistor. Any

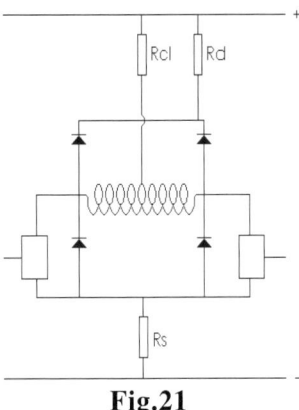

Fig.21

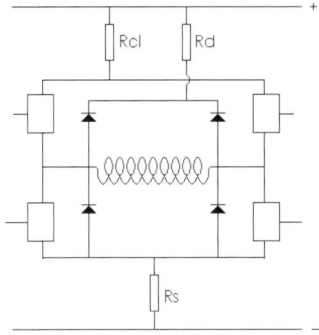

Fig.22

voltages through the sense resistor from the collapsing field can 'fool' some types of sensing circuit into thinking that the current limit as been reached before it has. This can have a detrimental effect on the torque of the motor. Some of the sophisticated drivers have a window area that ignores current sensing for a period greater than the likely length of the collapsing field. The other usual way is to connect the two diodes directly to ground. This affects the potential of the diodes relative to the ground potential of the driver. In practical terms the sense resistor is very small and connecting directly to ground makes minimal difference to the operation of the circuit.

The second problem area is around the current limiting resistor. This is often a large value compared to the resistance of the motor windings and the resistance of the current sense resistor. If this resistor also has an inductive element the current decay can be slowed to such an extent that damage can occur to the switching devices. Connecting the anodes directly to the positive rail can result in a high potential difference between the anodes of the diodes and the centre tap of the motor. A second much lower wattage resistor that does not have any inductive element can be used to connect the diodes to the positive rail. The calculation of this resistor depends on the voltage applied, the maximum current and the characteristics of the switching components. The type of current

control circuits being used can affect the worst case dissipation of this resistor.

Fig.21 shows a revised circuit for one coil pair of a unipolar drive. Rd is the diode resistor

Fig.22 shows a revised circuit for one coil of a bipolar drive.

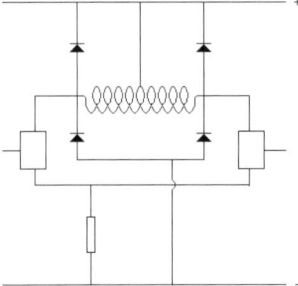

Fig.23 shows an effective practical circuit for one coil pair of a unipolar drive

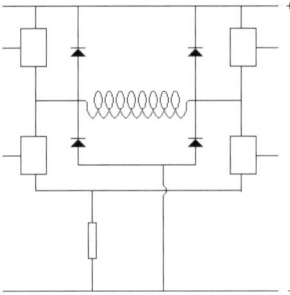

Fig.24 shows an effective practical circuit for one coil of a bipolar drive

The removal of the current limiting resistor and making the current sensing resistor as small as practicable will make the circuit more efficient but a current limiting circuit will be needed.

All the points above relate equally to the bipolar drive as to the unipolar drive.

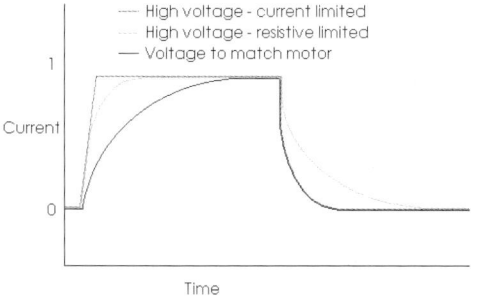

Fig.25 shows a graph of the rise and fall of current through a motor when using different voltages and different methods of current control. The fall of the current limited voltage will follow the plot for the motor matched voltage because the impedance of the motor is a fixed part of the circuit.

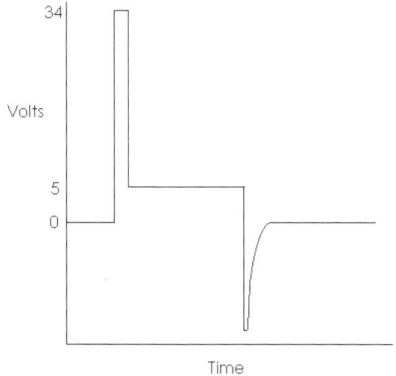

Fig.26 shows a graph of the rise and fall of voltage when using a switched voltage controlled by current sensing

The following circuit description is easy to implement using current sensing and voltage switching methods described in other sections of the book. A 12 – 0 – 12V transformer is connected for 24V. This is rectified and smoothed to give a DC voltage that will be 24 x √2 this approximates to 34V. At each motor step the voltage is set to this voltage and the current is sensed. When the current reaches the required level the voltage is dropped to the running voltage of the motor where the self-resistance of the motor will limit the current to the required level. This is shown on the graph as 5V but could be any voltage to match the motor. This occurs only once during each step and unlike the continuously running chopper circuits cause the minimum of switching interference.

The inverse spike shown is a result of the collapsing field and will be minimised by the diode network.

Recognising unknown stepper motors

When buying 'off the shelf' motors, winding data etc is available. Problems occur in the service industry or when stepper motors are obtained from scrapped units and have no manufacturers markings or are made for a specific task and the data sheet is not traceable.

The first step should always be to locate a data sheet on the particular motor or a similar type.

Voltage and current rating

This is not exact but once the specific type of motor is defined e.g. unipolar, the resistance of the windings can be measured. The frame size can be compared with manufacturers data and using the resistance data an educated estimate can be made of the phase current. The same data will normally indicate a voltage for the motor type, but if not this can be calculated from the resistance and phase current using Ohms law. Because many scrap stepper motors particularly of the small type will be made for computer or light industrial use the favoured voltages will be 5V and 12V. Stepper motors obtained from industrial units tend to have a wider range of voltages. Voltages from 3V through to 24V, 36V or 72V are not uncommon. The industrial motors also tend to be physically larger and often better labelled with manufacturers data.

Recognising windings

This is not as difficult as it may first appear because motors generally have a limited number of configurations. These are 4 wires, 5 wires, 6 wires and 8 wires. Simple measurements with a resistance meter will indicate the type and will indicate the different coils but not which end of the coils are which.

4 wire types

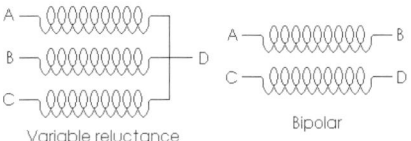

Fig.27 shows the two 4 wire possibilities. Simple metering will indicate if there are two separate windings or three windings with one common connection.
Two separate windings indicate a bipolar motor. Three windings with one common connection indicate a three-coil variable reluctance motor.

5 wire types

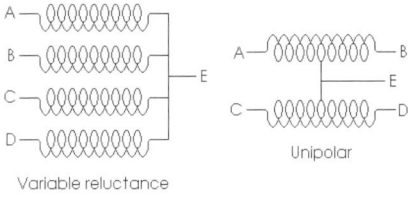

Fig.28 shows the two 5 wire possibilities. Simple metering will indicate the common connection. Four windings with one common connection indicate a four-coil variable reluctance motor or a centre tapped unipolar motor. The centre tap will be half the resistance from one coil end of the resistance between a full coil ends i.e. the resistance between A and E will be half that of A to B in a standard 'off the shelf' motor.

To distinguish between variable reluctance motors and centre tapped unipolar motors turn the rotor by hand. The variable reluctance motor will turn freely but the unipolar motor will move in distinct steps between the permanent magnet positions.

6 wire types

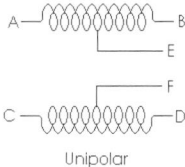

Fig.29 shows the layout of the most common possibility for a six wire motor. This is a unipolar motor with independent centre taps to the coils. It is possible to meet five coil variable reluctance motors but these are not very common. The same methodology applies to six wire as to five wire stepper motors. As previously, turning the rotor by hand will indicate if the type is a variable reluctance or a permanent magnet. This separate winding allowed the unipolar motor to be wired as a bipolar motor by ignoring the centre taps.

8 wire types

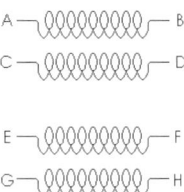

Fig.30 shows the layout of an eight-wire stepper motor. This type of motor can be configured as a series pair bipolar, a parallel pair bipolar and a unipolar motor.

Step recognition

This is more difficult than winding recognition

Inside of a small stepper motor showing coils and 'cog' configuration.

but a few logical steps will enable even the most complex winding to be sorted out.

The bipolar winding can be connected to a suitable bipolar driver. If the step direction does not match the notional direction of forward and reverse, reverse one of the pair of windings. The approach to the variable reluctance irrespective of the number of coils is the same as for a unipolar motor. Use a suitable size battery to match the voltage of the motor. Attach the common point to the positive of the battery. Select a notional 'A' wire and touch it to the negative of the battery and remove, the motor will jump to a pole position. Touch one of the other leads to the negative of the battery and remove. If the motor moves CCW, the connection is wrong, touch 'A' again to the battery and select another wire to touch. The possibilities are a step CCW, a small step CW and a double step forward depending of course on the number of coils. The object is to find the lead that produces a small step CW. This becomes lead 'B'. Ignore lead 'A' and use 'B' as a starting point. Find the next lead that produces a small step CW from 'B'. This will be lead 'C'. Continue until all the leads are labelled. The CW operating sequence is now 'A', 'B', 'C' etc and the CCW direction will be the reverse of the sequence from any of the pole position i.e. if the motor was at pole position 'B', CW would be a next step of 'C' but CCW would be a next step of 'A'.

Eight wire motors present a particular problem because of the number of permutations that can occur. Although there is no standardisation, some manufacturers seem to be trying to make recognition a little easier. Some eight wire motors do not need a meter for coil recognition because coil wires are colour paired e.g. red and red and white stripes. The striped wires are intended as the centre taps. This does not of course indicate the correct phase. If all the striped wires are assumed to be common for unipolar configuration, the same procedure can be used to find the phase sequence with a battery as previously. Once the unipolar phase order is known this can be translated to bipolar connection if required.

If a wrong connection is made to a driver, provided it does not produce a power to ground short, the result is rarely catastrophic. Fuses placed in the circuit during testing should prevent major problems.

If a stepper motor is connected incorrectly it is usually obvious because the stepping will 'dither' around one position or will appear erratic often appearing to be a series of mixed short and long steps.

Non standard stepper motors

Problems occur when trying to replace a motor, or use a motor that is from a different application. When an OEM is buying motors in hundreds of thousands, the motor is often non-standard to make manufacture of other adjoining parts easier or in some cases to make servicing by a third party more difficult.

'Strange' meter readings can be indicative of extra components such as diodes or resistors built into the body of the motor. This needs to be taken into account when replacing the motor or using the motor in a different application.

Practical stepper motor drive circuits
Using the L297

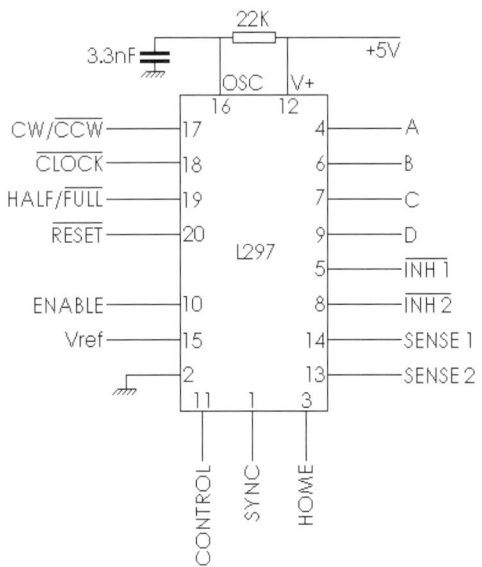

Fig.31 shows the L297 layout. This is a simple IC to use with the exception of drive method choice. The L297 can be used with dedicated drivers such as the L298 or with discrete drivers for higher current use.

A simple capacitor resistor oscillator is based around pin 16. The frequency approximates to 1/0.69RC. There is a SYNC input output on pin 1 that allows one IC to be set up as master and the oscillator frequency to passed on to other L297s being used. This ensures that switching occurs in synchronisation and helps to reduce ground noise. If the multiple oscillator option is used pin 16 of the slave devices is tied to ground.

An ENABLE output is held high but can be taken low to force low INH 1, INH 2, A, B, C and D low.

There are two SENSE inputs, one for each 'H' bridge, that are used to measure the voltage developed across resistors. These SENSE inputs are used in conjunction with the Vref input that can have an input between 0V and a maximum of 3V. When the voltage developed across the sense resistor is equal to the voltage on Vref the current control flip-flop is reset therefore chopping the current. The flip-flop is reset on the next oscillator cycle and is chopped again when the current again reaches the peak set. Consider a requirement for a maximum current of 2A with sense resistors of 0.5W. When the current through the resistor reaches 2A the voltage across it by Ohms law equals 1V. Therefore if the Vref input is set to 1V the current will chop at 2A. The INH inhibit signals are intended for use with bipolar drivers such as the L298. They allow chopping to be used with the enable inputs instead of the phase lines. There is an advantage with the L298 in that dissipation is reduced. With the CONTROL pin high chopping occurs on the phase lines and when the CONTROL pin is low chopping occurs on the inhibit lines. The phase lines denoted A, B, C and D are pins 4,6,7 and 9 in that order.

The most complicated part of using the L297 is the selection of drive methods. This is based around a section of the IC called the translator. **Fig.32** shows the output of the translator in box form. The cycle increments when a low to high transition occurs on the CLOCK input. Direction is chose by the CW/CCW pin. This is chosen by a high or a low on this pin but the actual direction of motor movement depends on the

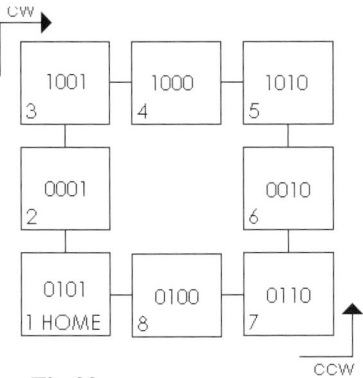

Fig.32

connection of the motor windings.

A low on the RESET pin moves the translator to box 1, the home position. An open collector output on the HOME pin indicates that the cycle is at the home position.

There are three possible drive modes - half step, normal drive and wave drive.

Half step is selected by a high on the HALF/FULL pin. The step sequence is through all boxes alternating between one coil being activated and two coils being activated.

Normal drive or two phase on drive is selected by a low on the HALF/FULL pin when the translator is at odd numbered states i.e. 1, 3, 5 or 7. The step sequence is through all the odd boxes where two coils are activated.

Wave drive or one phase on drive is selected by a low on the HALF/FULL pin when the translator is at even numbered states i.e. 2, 4, 6 or 8. The step sequence is through all the even boxes where one coil is activated.

A feasible start up sequence is to put a high on the HALF/FULL pin thereby selecting half step mode. Put a low on the RESET pin to move the translator to the home position.

The home position is an odd position therefore putting a low on the HALF/FULL pin will set up the normal or two phase on drive.

If wave drive is required it is necessary to move to the home position and make one step to an even position and put a low on the HALF/FULL pin at this position.

In practical terms it is sometimes necessary to change drive modes during operation when it is not practical to return to the home position. This is easily achieved particularly if a computer is being used as the controller. It is necessary to use a two bit counter that changes state with each CLOCK input. The HOME output can be used for synchronisation. A flip-flop can be used with the outputs toggling between states at each clock pulse and the HOME output being used as a reset to synchronise the count.

Using the L297 with the L298

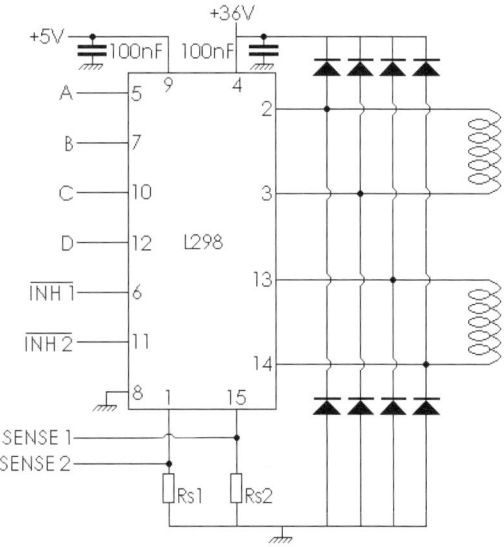

Fig.33 shows the L298 set up as a bipolar driver for connection to the L297. This device can work with voltages to a maximum of 46V and coil currents of 2A per coil. The L298 requires a logic level voltage on pin 9.

For clarity the circuit is shown with wire jump and marked connections.

The diodes need to be high speed and rated for a minimum of 2A.

The L297 can also be used in other 'H' bridge driver circuits to a maximum of 4A total current.

Pin designation for BM101P		
Pin	Pin name	Description
1	Stop	Sets all motor outputs to zero. Can be used as an emergency
2	CS	Chip select, active positive, tie to + rail if not used. If
3	CW/CCW	Direction selection. Tie to – rail for CW or tie to + rail for CCW
4	Step	Steps one count on positive going clock input.
5	V_{SS}	Zero or ground connection.
6	Drive 1	Output 1 – see connection diagram.
7	Drive 2	Output 2 – see connection diagram.
8	Drive 3	Output 3 – see connection diagram.
9	Drive 4	Output 4 – see connection diagram.
10	Drive 5	Output 5 – see connection diagram.
11	Drive 6	Output 6 – see connection diagram.
12	Drive 7	Output 7 – see connection diagram.
13	Drive 8	Output 8 – see connection diagram.
14	V_{DD}	Positive rail connection *1
15	Full step*2	Tie to + rail to select otherwise tie to - rail
16	Half step*2	Tie to + rail to select otherwise tie to - rail
17	Wave*2	Tie to + rail to select otherwise tie to - rail
18	Bi/unipolar	For bipolar tie to + rail to select, for unipolar tie to - rail

*1 *The positive voltage range is +3V to + 5.5V, but it is recommended that a regulated +5V power source is used.*
Maximum input is V_{DD}. Attenuation must be used with inputs exceeding this value. Over voltage input protection is recommended because the device may be damaged with voltages exceeding V_{DD}.
*2 *If more than one of these tied to + rail priority will select in order of full step, half step and wave.*

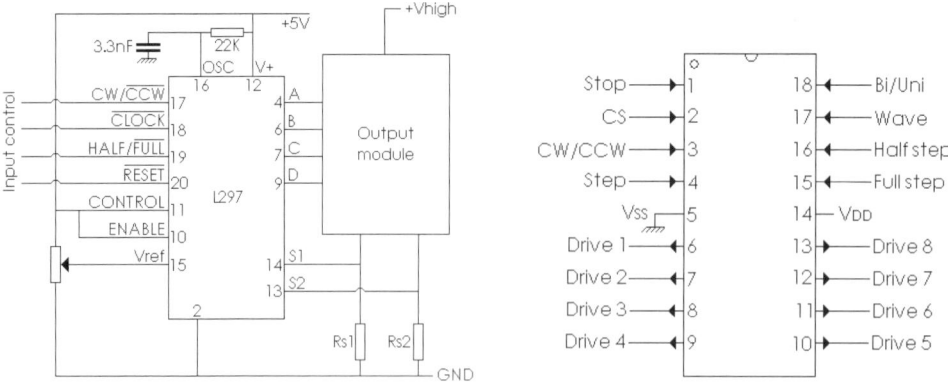

Fig.34   Fig.35

Electromechanical Building Blocks

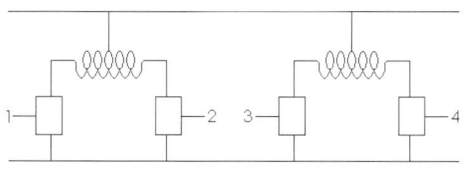

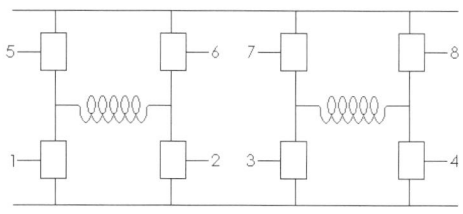

Fig.36 **Fig.37**

Fig.34 shows a circuit for using the L297 to drive a number of high power output modules. The input controls can be from a PC or from a manual controller. The drive mode selection can be manual using the count flip-flop described earlier as a guide to odd or even position or can be from a sequencer that runs at start up.

If Rs1 and Rs2 are 0.1W resistors then 10A will give a voltage drop of 1V across the resistor. Commonly available stepper motors rarely exceed this level of current. The Vref can be set to produce chopping theoretically up to about 30A. The heat dissipation from the resistor would be high therefore for high currents it is more feasible to parallel two resistors. Stepper motors that use currents at very high levels are likely to be used for specialist applications and are probably better driven by an IC or a microprocessor designed specifically for the application that as the facility of voltage switching to lower the heat produced during hold times.

The BM101P microprocessor bipolar or unipolar stepper motor driver

Fig.35 shows the BM101P layout.

General description

The BM101P is a driver for bipolar or unipolar stepper motors. It has a number of selectable drive features. The IC is intended for the driving of transistor or MOSFET type 'H' bridge circuits. The IC has the following features. A stop feature that sets all motor outputs to zero, or can be used as an emergency stop, or as a count sequence reset.

Bipolar or unipolar drive and full step, half step and wave drive is selected by jumper or taking pin to power rails.

Fig.36 shows the drive pin designation for unipolar drive.

Fig.37 shows the drive pin designation for bipolar drive.

Microstepping

Microstepping is a technique for balancing the current in the coils of a stepper motor to give intermediate steps between normal full or half step positions. One of the problems of running a stepper motor at slow step speed is the mechanical noise and vibration due to the inertia of the rotor in start/ stop operation and 'ringing' in the power circuits due to the inductance of the windings. Microstepping can reduce these effects to very low levels by using true or step simulated sine/ cosine current waveforms. Unfortunately microstepping is not as accurate in terms of absolute positioning, resolution and repeatability as some driver unit advertising implies.

All stepper motors have inherent errors that are usually expressed in terms of a full revolution or +/- a full step tooth position. This error is not cumulative but occurs independently at each step. The errors occur because of many factors but include friction, magnetic hysterisis, machining tolerances and general design such as tooth shape of the rotor and stator.

A standard unipolar/ bipolar driver will work with a wide range of stepper motors to produce steps within the manufacturers tolerance.

Microstepping drivers on the other hand require

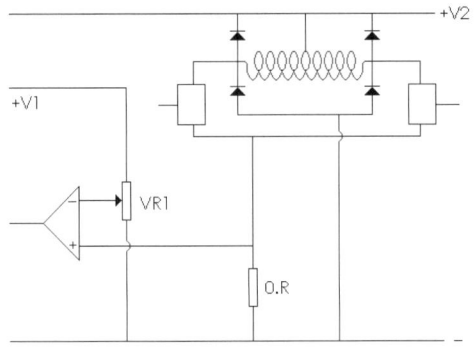

Fig.38   **Fig.39**

designing to match the characteristics of the motor type to produce high accuracy.
Microstepping is claimed as a method of cutting out the need for gearing with a stepper motor. Consider the following example using information from a manufacturers data sheet.
Step angle = 7° 30'   Step angle tolerance = +/- 40'
If a 1/32 micro-step was chosen this is equivalent to 14'.
The step angle tolerance of 40' is 2.86 microstepping positions. With microstepping units giving 1/64 step this is an error of 5.72 microstepping positions. Because this could be either side of the theoretical the error would be +/- 5.72 microstepping positions. This is 11.44 microstepping positions of 14' which is 2° 40' on a 7° 30' step. This approximates to a 35% non-cumulative error.

With a geared down system the step angle and the step angle tolerance are both reduced by the gearing factor but can of course introduce problems of their own e.g. drag, and tooth play if used in two directions. Toothed belt synchronous drives are often used in positioning critical applications. These have more drag but do not have the problem of tooth play. The system torque is increased but the system speed is reduced. Therefore the direct load on the motor may be reduced. This may allow faster motor stepping that can to some extent make up for the slower system speed.

Current control for stepper motors

Current limiting can be easily achieved for stepper motors. If the current is monitored it is possible to control the V2 voltage or to cause shut down of the driver circuit. The V2 voltage can be from a multi voltage supply.

Fig.38 shows a driver circuit for a unipolar motor with a very low value resistor in the ground circuit. A voltage comparator is used to measure the voltage developed across the resistor and hence the current passing through it. This voltage is compared with the voltage on the threshold voltage from VR1. Changing the +/- inputs over will cause a falling instead of a rising output. The circuit will trigger on the peak current taken by the stepper motor coil.
This circuit can be used to flag current set level to a microprocessor controller.
A method often used with peak current sensing is to use the output of the comparator to trigger a monostable that inhibits the output driver. At the end of the monostable timing cycle the output driver switches back on and the cycle is repeated each time the current limit is reached. The drive is therefore in the form of a chopped output. This type of circuit can be set up independent of the

controller by using and gate type logic on the inputs of driver components. Or the circuit can be used to switch to a lower V2 level in designs using multi level voltages for the stepper supply.

Practical stepper motor MOSFET drive circuits
Fig.39 shows a practical 'N' channel MOSFET 'H' bridge drive for half of a bipolar stepper motor. The high side drive is from a boost voltage. Fet1 is a small MOSFET such as the 2N7000. This is capable of switching 60V drain to source with a drain current of 200mA. It comes in a TO92 package. Fet1 is switched from an inverting buffer. With no input the inverter switches on Fet1 and R2 is pulled to ground. The junction of R1 and R2 is +ve boost multiplied by the ratio of the resistors. Suitable starting values are 10K and 4.7K for R1 and R2 respectively. This provides a gate voltage of about one third +ve boost. With a positive input to the inverting buffer Fet1 is switched off and the junction of R1 and R2 rises to +ve boost. The circuit is duplicated for the right hand drivers. Suitable gates are 4011 for the 2 input nand gate and 4049 for the inverting buffer. A suitable power supply must be provided for the CMOS logic. The gate and gate to source resistors are as described in the theoretical section on the MOSFET 'H' bridge. Current chopping is applied to one input of both low side drivers consecutively. Only one low side driver will be selected. The small circuits shown can replace the 2 in nand gate if current sensing is not required.

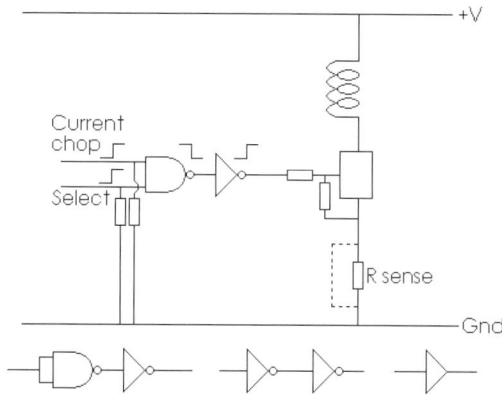

Fig.40 shows a section of a practical 'N' channel MOSFET drive for one coil of a unipolar stepper motor. The circuit is repeated for each coil. With this circuit it is advantageous if current chopping is used to apply it to all coils individually using the current control circuits described earlier. The small circuits shown can replace the 2 input nand gate if current chopping is not required. The non-inverting buffer is a 4050.

CHAPTER FOUR

DC Motor Drive

There are many small DC motors available at a reasonable price but unfortunately the price seemed to increase out of proportion when higher power is required.

Types of DC motor

Small DC motors with the exception of multi coil electronic switched motors consist of two parts. The first is a fixed magnetic field that is usually a permanent magnet but can be a field coil wound around an iron pole piece called the stator. The second part is a set rotating coils wound around a former. The connections are brought out to a segmented ring around the shaft called the commutator. Two 'brushes' rubbing along opposite sides of the commutator provide power and the sequence of switching of the rotor coils.

Motor power

Choosing a suitable DC motor for a project is often a matter of experience, guesswork and trial and error.

Many manufacturers quote current consumption without stating the load conditions this was measured at. The input power quoted of a small DC motor does not necessarily reflect the usable available output power under 'normal' conditions. The better manufacturers and suppliers will usually provide power/ load graphs and efficiency figures for their motors.

A middle of the range permanent magnet DC motor may be only 25% efficient. This means that for every 100W of electrical energy put in only 25W of mechanical energy is available at the output.

Most types of household equipment with motors fitted are usually quoted in the form of input energy therefore most people are familiar with comparing from these type of figures. The problem starts when comparing and choosing equipment on this basis with no knowledge of the basis of the measurement or the true efficiency under different load conditions.

Field wound motors

These follow the same fundamental rules as permanent magnet motors except that the fixed magnetic field is produced by current flowing in a fixed outer coil or coils around the motor instead of by permanent magnets. This coil is called the field or stator winding. These motors rotate in a fixed direction irrespective of the polarity i.e. it is not possible to reverse direction of the motor by reversing the power leads. This is because of the current movement relationship and hence the magnetic fields produced in the stator windings relative to the rotor windings. Changing the relative current flow in either the field or rotor winding, but not both, can change the direction of rotation.

With a few minor modifications and provisos these types of motors can be treated as permanent magnet motors for the purpose of controlling speed and direction.

They are often less expensive to produce for motors needing a relatively high output than the comparative output permanent magnet motor that often uses expensive rare earth materials for the magnets.

Series wound motors can be used with both AC and DC power. Theoretically parallel wound motors can be used with both AC and DC power but because the field windings are directly across the supply they are usually of thin wire with a large number of turns. This produces a high DC resistance and a high inductance. The result when using AC is a shift in the current peaks in the field winding compared to the rotor windings and hence low output power and inefficiency.

In a low voltage DC form these motors are often used where there is a low voltage low impedance i.e. high current power source, e.g. motor vehicles.

Because of the simple construction of these types of motor the wiring is simple to modify. The rotor windings can be traced from the brush housings and the field winding connections are from each end of the field winding. Simple metering will show if the motor is series or parallel wound.

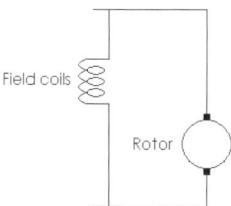

Fig.41 shows the connections for a parallel connected motor.

Series wound motors

As the name implies the field winding is wired in series with the rotor winding. This generally does not provide as much power or as good speed regulation as the parallel wound motor. This is because the input voltage is shared between the field and rotor winding and the varying back EMF of the rotor winding under load will affect the current flow in both the stator and rotor windings. These motors rotate in a fixed direction irrespective of the polarity i.e. it is not possible to reverse direction of the motor by reversing the power leads. This is because of the current movement relationship and hence the magnetic fields produced in the stator windings relative to the rotor windings. Changing the relative current flow in either the field or rotor winding, but not both, can change the direction of rotation.

As with the permanent magnet the motor power

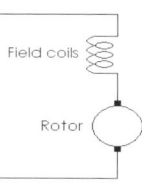

and speed is relative to the input voltage.
Fig.42 shows the connections for a series wound motor.

Compound wound motors

As the name suggests the compound motor is a combination of the series wound motor and the parallel wound motor. The characteristics depend on the actual windings and the winding relationship between the parallel winding and the serial winding.

The two parallel windings act in a way that they both reinforce the field magnetism. If it is necessary to change direction by changing the relative current flow between the field coils and rotor then both field coils must be changed over. Changing just one coil over will result in a weakening of the field magnetism. Direction change may or may not occur depending on the relationship of the partial magnetic field due to each of the field coils.

Fig.43 shows the connections for a compound wound motor.

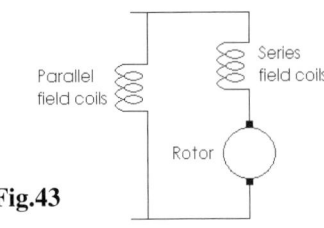

Fig.43

Comparative characteristics of series, field and compound wound motors

The behaviour of the series wound motor and the parallel wound motor are totally different. The compound wound motor falls somewhere between the two depending on the design of the windings.

Fig.44 shows a graph of the relative torque

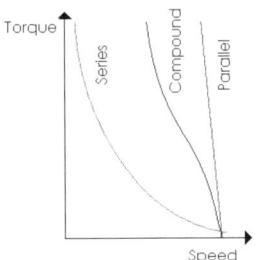

speed outputs for each type of motor with a constant supply voltage.

The series motor produces a high torque at low speed. This is because the low back EMF generated allows large currents to flow. When the speed increases so does the back EMF and hence the current through the windings falls with a reduction in current and hence torque. This characteristic can be useful for driving units requiring a large starting torque to begin movement such as battery driven vehicles. With a small motor the losses, friction and inefficiencies inherent in the motor will limit the top speed of rotation. With a large motor this is likely to be a much lower percentage of the motor power and therefore with no load, very high speeds are possible. The rotational forces produced may be sufficient to cause the armature to 'burst'.

The parallel motor is in effect a constant speed machine for a given supply voltage. Any variation in load will be counteracted by a change in speed and hence a change in the back EMF that will cause a change in the current through the armature. The current in the field will remain constant. The effect is that the motor will attempt to maintain a constant speed. Losses in the windings, friction and inefficiencies will prevent the motor achieving constant speed characteristics hence the graph is a slope approaching vertical.

The compound motor is designed to have characteristics between the series and the parallel motor.

The exact shape of the graph will depend on the design of the individual motor.

When using electronic control with motors the maximum current likely to be taken by the motor must be used in determining the rating of the driver components. Little damage may be caused to a motor by exceeding the current or voltage specifications for a short time. Exceeding the current and voltage ratings of many electronic components will result in irreversible damage. Care must be taken particularly when choosing driver components for series wound motors because of the high starting current of this type of motor.

Using series, field and compound wound motors
Parallel wound

As the name implies the field winding is wired in parallel with the rotor winding. Unlike the series wound motor the field current is and hence the stator magnetic field is constant for a given voltage supply.

As with the permanent magnet the input voltage controls motor power and speed. But unlike either the permanent magnet motor or the series wound motor both the field and rotor windings can be supplied with a voltage independently of each other. Voltage change in either or both will have an effect on speed and power.

If both the field and rotor windings are used in parallel with a PWM supply the effect can be inefficiency due to the high inductance of the field winding reacting to the changing voltage of the PWM as if it were an AC voltage. The answer to this can be to use a constant field voltage and use PWM on the rotor winding only. The field winding can theoretically be used for speed control but is rarely used in practical situations. Usually the field winding is run near the point of magnetic saturation and therefore the only control that can be used is a decrease in the field current and hence the magnetic field. This is not a very efficient way of using this type of motor. Direction change can be achieved by changing over the connections to the rotor winding or by changing over the polarity of the field winding. Simple circuits can be made that switch on the field winding only when a PWM signal is applied to the rotor winding. A relay or 'H' bridge circuit can be used to control the current flow direction in the field winding and therefore the motor direction.

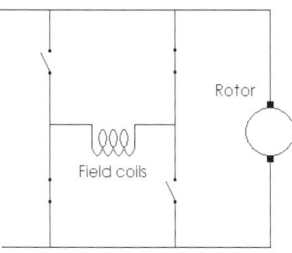

Fig.45 shows the connections for a parallel connected motor using a change over relay to provide reverse rotation. An 'H' bridge circuit can be substituted for the change over relay.

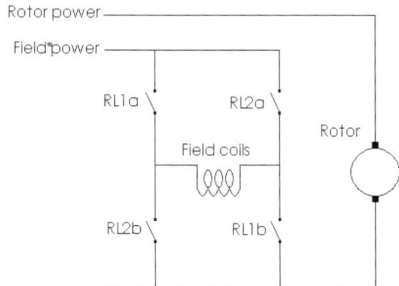

Fig.46 shows the connections for a parallel connected motor using two relays to provide reverse rotation. Using two relays allows disconnection of the field coils when the motor is not running. It is important that the field coils are not disconnected with the rotor coils in circuit. This will lead to burn out of the rotor coils. A change over relay could also be used in this circuit if one direction was predominant. The circuit could also be modified for use with 'H' bridge circuits.

Fig.47 shows a layout for automatically

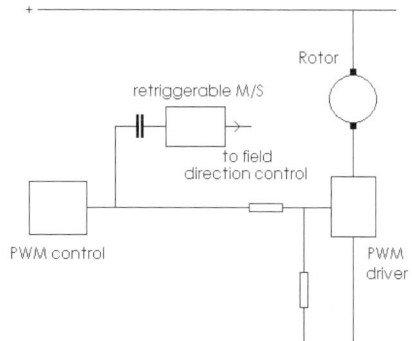

providing field coil current when the PWM triggers. A positive triggered retriggerable monostable is used that as an on time longer than the frame time of the PWM. This means that the monostable will begin timing at each rise of the PWM pulse and remain on longer than the time to the next potential pulse from the PWM irrespective of the PWM on/ off ratio.

Direction changing by input polarity change

It was stated earlier that the direction of rotation of a series wound motor was dependent on the relative current flows in the windings and did not change if the input voltage was reversed. If the field winding current is supplied from a bridge rectifier the current flow will be in a constant direction independent of supply polarity. The rotor current flow will change direction with the input voltage polarity. Therefore the motor will change direction with change of supply polarity. The bridge rectifier needs to be of a large enough current capacity to supply both windings.
A similar technique can be used with parallel

wound motors that are not intended for PWM control. The bridge rectifier in this case carries only the current for the field winding.

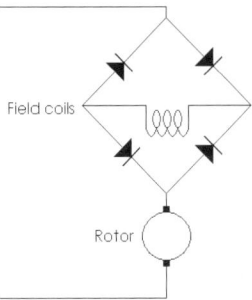

Fig.48 shows the connections for a series connected motor using a bridge rectifier to maintain a constant direction of field current irrespective of power connection polarity. Connecting the rotor via a bridge rectifier instead of the field coils will have the same effect but the rotor current direction will remain constant with changes of polarity.

A range of DC motor brushes. The plate assembly consists of permanently mounted phosphor bronze springs with attached contact point. Failure of the motor was due to breaking of a contact point.

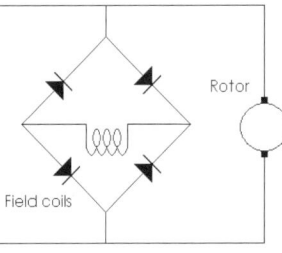

Fig.49 shows the connections for a parallel connected motor using a bridge rectifier to maintain a constant direction of field current irrespective of power connection polarity. Connecting the rotor via a bridge rectifier instead of the field coils will have the same effect but the rotor current direction will remain constant with changes of polarity.

Compound wound

The circuits shown for the other two motors can be modified for use with the compound wound motor but the field coils must be treated as a single entity and remain in phase. It is easier to make changes around the rotor.

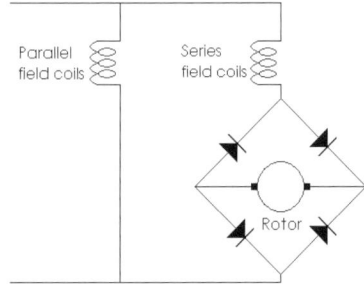

Fig.50 shows the connections for a compound connected motor using a bridge rectifier to maintain a constant direction of rotor current irrespective of power connection polarity.

Permanent Magnet Motors

Brush motors

This type of motor can be regarded as a parallel field coil or shunt motor. The field windings have in effect been replaced by a constant magnetic

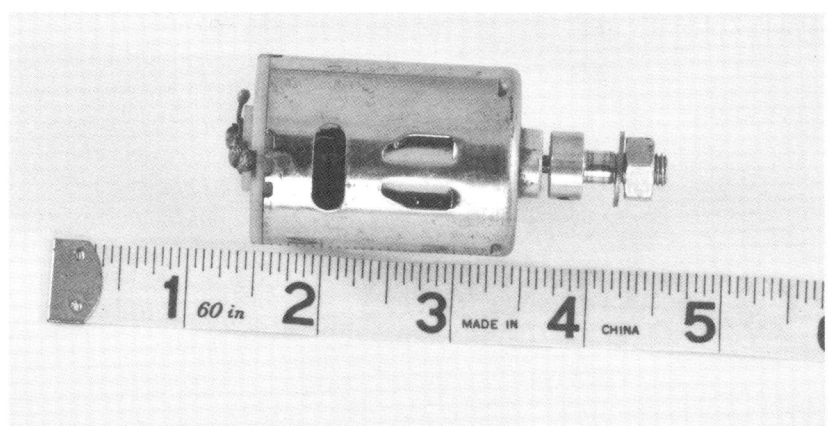

A 400 type motor

field provided by permanent magnets instead of a coil.

The number of magnetic poles around the fixed field and the moving field is usually an offset number to prevent alignment e.g. the rotating field is often in multiples of three and the fixed field in multiples of two. As the rotating section rotates a brush system on the commutator switches the coils so that continuous movement occurs. The brushes are normally made of a graphite composition but in small motors may be a phosphor bronze or copper strip.

If a motor is spun without a voltage being connected, the motor will act as a generator and produce a voltage that is called the back electro motive force, abbreviated to back EMF. This voltage is still produced when power is applied to the motor and the motor will increase in speed until the back EMF balances the input voltage. The applied voltage therefore affects the speed directly.

Brushless motors

The brushless motor works on the same principle as the brush motor but the rotor in this case is made up of a number of fixed magnetic fields. The stator consists of the coils. The switching of power to individual coils is done electronically and position sensors are used for switching control. Many brushless motors are used in low power drives. They are often used in computer equipment because they do not produce interference as a result of arcing brushes.

Outrunner motors

These are the big brothers of the brushless motor. The larger of these motors are capable of producing outputs in the order of 4KW – over 5HP. They are physically very small and mainly designed for short duration runs and unfortunately have price tags to match their design along with needing controllers that have a similar price tag.

A DC motor brush layout

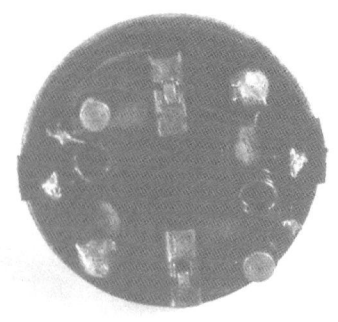
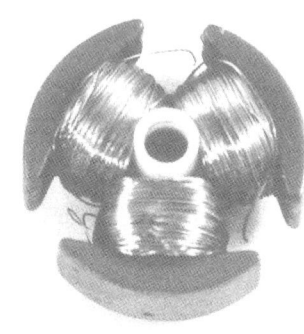

A complete rotor and brush assembly of a five pole motor showing wear on the commutator. Although a two pole motor is possible, simple commercial designs favour a minimum three pole configuration.

The design consists of an outer rotating drum to which is fixed a number of permanent magnets. The stator is a number of coils around pole pieces. One of the common designs is fourteen magnets and six or twelve coils. The coils are wound in three phase format i.e. the end of one set of windings is attached to the start of the next set of windings. This gives only a three-wire output unlike the other motor that has two ends for each coil or the 3-pole variable reluctance motor that has three coil ends and a common point. Coils are switched in sequence by an electronic controller and the switching rate is a factor in the speed control. The motors and controllers are normally matched for a particular application.

Low power motors

These motors turn up almost everywhere from toys to household goods. They are available from many sources. They are designed for a wide range of voltages often as low as 1.5V.

Higher power motors

Higher power motors are relatively expensive. I needed a number of small powerful motors that could be easily controlled both in terms of speed and direction and would be available at a reasonable price.

I looked at a number of other sources including buying a rechargeable drill and stripping out the motor. This was not an inexpensive option, but was less expensive than purchasing similar power motors from the normal range of suppliers. I had been given a 'slightly' beyond repair radio control car – just in case any of the parts 'may be useful'. This had sat in a corner of my workshop as another 'get round to it at some point' job. But I knew that it was battery powered and therefore worthy of some investigation.

The car was fitted with a small powerful motor that was intended for use with a battery pack of 7.2V. Tests on this motor showed that it would run for a reasonable length of time at 12 volts before becoming hot to the touch. Stall tests immediately blew the 10A fuse in the battery charger being used. Substitution of a larger power supply gave a stall current of 14A. This is an input of 168 watts. This is almost a quarter of horsepower input from a motor of less than 1.5" in diameter and 3" long overall. It would not of course be wise to run the motor at this level and when the motor is stalled it is obviously not producing any drive only heat! One manufacturers specification for this type of motor gave a stall current in the order of 21A and an operating voltage of 3.6 to 8.4 therefore there was a discrepancy between this and my results. This could be because of the individual motor, or it could be from a different manufacturer because there were no markings on the motor or as often happens with data tables they may not contain the full test criteria and the stall test may have been at a different voltage level using a constant voltage supply.

A visit to a model shop showed a range of motors. Then a visit to a newsagent produced a magazine on electric powered flying models, the advert pages were full of motor and small gearbox options. Finally a www visit revealed even more exotic brushless motors capable of input powers in the order of 4KW – over 5HP! These were physically very small and mainly designed for short duration runs and unfortunately had price tags to match their design along with needing controllers that had a similar price tag. The efficiency of some of these types of motor is claimed to be typically 95%. Returning to the real world there is a range of motors classified as type 280 to 900. These are available in a variety of voltages and power outputs. The motor fitted to the car was of a size that seemed to match the specifications for a 400 type, this is a lower to middle range motor in terms of its power and specification and a replacement cost in the order of £5. At this kind of price further investigation was justified.

The motors

The specifications for the motors were slightly confusing especially when comparing the test reports. The tests were often carried out relative to the number of battery cells. Motors were either designated by voltage or sometimes in the form of cells e.g. 8/10 cells. Most small motors used 7.2, 8.4, 9.6 or 12V. The cell designation was based on NiCd cell voltage of 1.2 volts i.e. a 10 cell battery is 12V.

Many of the motors were overrun, particularly for use in aircraft. This did not seem to harm the motor, probably because flight times are relatively short, full power is usually only used at take off or during climbing manoeuvres and there is a good flow of cooling air.

Taking all this into account the motors still appeared a good proposition for a number of projects even with the motors running within their ratings. It is necessary to use some form of control to adjust the speed and the first option is gearing, this will also increase the torque of the motor. A number of gearboxes are available inexpensively for these motors. There are gear drive, belt drive and planetary versions. Unfortunately the ratios are limited and are based on propeller size and pitch. The maximum ratio commonly found is 3:1. This is still a good starting point because it provides a mount for the motor and a larger diameter output shaft. Most electromechanical devices such as power relays and solenoids are based on operating voltages of 12 or 24V. This would therefore need to be the voltages that the motors would run at. It is simpler to make the motor conform to the power supply than regulate the power supply to match the motor particularly at the current levels envisaged.

Pulse width modulation was the answer because it can be used to control the running speed, and also maximum and minimum power limits of the motor can be easily set.

The following table is a précis of information I gleaned from various sources and gives an

Motor	Nominal voltage	Voltage lower	Voltage upper	No load RPM	No load current	Watts at no load V1	Watts at no load V2	Current @ max efficiency	Watts @ max efficiency V1	Watts @ max efficiency V2	Stall current
280	6	4.5	6	14000	0.3	1	2	1.6	7	9	6.8
300	6	1.2	6	29000	0.7	1	4	5	6	30	28
400	6	2.4	7.2	18000	0.7	2	5	4	10	29	25
400	7.2	3.6	8.4	16400	0.5	2	4	3.3	12	28	21
480	7.2	7.2	9	17000	1.1	8	10	5	36	45	31
500	7.2	3.6	8.4	21200	2	7	17	14	50	118	96
500E	12	6	12	12000	0.4	2	5	2	12	24	10
600	7.2	3.6	8.4	18200	2	7	17	12	43	101	85
600	8.4	4.8	9.6	15500	1.8	9	17	11	53	106	70
600	12	4.8	14.4	17200	1	5	14	7	34	101	40
700	9.6	4.4	14.4	15000	2	9	29	12.5	55	180	65
700	12	7.2	19.2	11600	2	14	38	12.5	90	240	43
820	20	16	24	24000	2.5	40	60	21.3	341	511	166
900	12	6	40	6500	1.1	7	44	8	48	320	54

indication of the motor specifications but the information should only be regarded as a starting point. The motor type applies mainly to the physical size of the body. A 400 type motor may be available with many different number of winding turns and different thickness of wire for the windings. These factors affect the current rating and hence the power developed and may also affect the run time before the motor reaches a high temperature.

Once the requirements for a motor of this type to be used in a project are decided on, the next best step is to visit individual manufacturers web pages or to visit a good model shop.

The motors vary in body size depending on type specification physically from about 28mm diameter by 30mm length to about 50mm diameter by 95mm length. The shaft lengths vary also and shaft diameters range from 2mm to 6mm.

Motor speed and direction control
PWM – pulse width modulation

This is a technique to control the speed of a brush type DC motor by varying the time the power is switched on for within a set time frame. If the power is switched on for 100% of the time, the motor will run at full speed. If the motor is only switched on 50% of the time the motor will run at approximately half speed. Approximately is used, because the actual speed can be affected by the motor characteristics and the load placed on the motor. The width of the pulse is variable from 0% to 100%. It is likely because of the starting inertia of the motor that the usable range will be in the order of 25% to 100%. PWM unlike other forms of speed control uses the full voltage at all times, it is only the time period of switch on that varies. Using digital switching means that the dissipation is low in the output components because the output driver is either fully on or fully off. These states give a theoretical zero dissipation, but in reality the output devices have a transition time between states and a small resistance that causes heat to be produced.

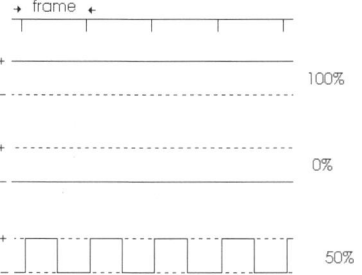

Fig.51 shows frame time compared to PWM output.

Motor speed control

It is possible to build a pulse width modulation speed control using discrete components. The circuit shown is typical of the type of circuits used. This shows a voltage ramp generator connected to the + input of a voltage comparator. When the voltage on the ramp exceeds the voltage level on the − input, the comparator switches on. It switches off when the input voltage falls below − input. It is difficult to design a linear ramp generator using analogue techniques because they are normally based around the charging rate of a capacitor. This charge rate is exponential. Constant current charging circuits can be used to make the charging rate more linear but in practical terms only work well at relatively slow charging rates. The optimum frame rate for most small motors is about 20 KHz. This is outside the human audio range. At slower speeds the motor can 'whistle' at certain speed settings. Most ramp generators do not work 'rail to rail', typically for ramp generators based on the 555 type IC, the voltage swing will be $1/3$ to $2/3$ rail potential. This will necessitate extra resistors in the − input of the comparator if the full range of the potentiometer is required.

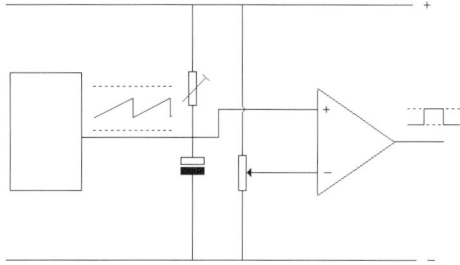

Fig.52 shows a block diagram of a simple PWM control.

A better method of control is to use a dedicated microprocessor. The microprocessor described contains a very accurate PWM based on digital timers. Using an ADC function to read a potentiometer, a 10-bit output (1024 discreet steps) can be used to control the PWM. If the PWM output is used to drive one or more MOSFETs, the control of high power DC motors is easy. It is feasible to start the motor with a pulse of 100% modulation to give the motor a 'kick' start then revert to the set speed. The microprocessor is capable of carrying out this operation at a speed that is undetectable to the observer.

The microprocessor is capable of combining both PWM and direction control using the same potentiometer, i.e. Half of the count is forward direction and the other part of the count can be programmed to give a reverse direction. A centre off band can be built in. Or a separate switch can be used to set the direction and the microprocessor program can be used to prevent the motor reversing until the speed drops to zero. It is also possible to develop microprocessor circuits that can be used to measure the speed of the motor and control the PWM width to stabilise speed with varying loads if a speed critical application is being undertaken.

PTM – pulse time modulation

This is a technique used in a similar manner to PWM but instead of using a variable pulse length as in PWM, a fixed pulse length is used and the time between pulses is varied.

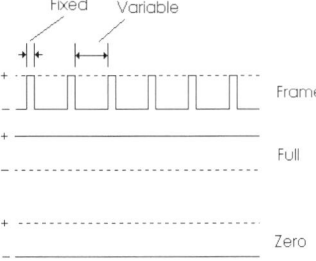

Fig.53 shows the waveforms of a PTM output. When the time between the pulses is zero the output is full. Because of the fixed pulse length zero cannot be achieved by using the variable time between pulses and a secondary 'blocking' gate is used for zero. The system has a number of advantages over PWM when a basic control system is required. The system is digital and can be used with systems not reliant on the charging characteristics of a capacitor although in the simple system a 555 type IC is often used that uses a capacitor for its time period. Set up is easier in that minimum speed offsets are not required, this is taken care of by the fixed pulse length.

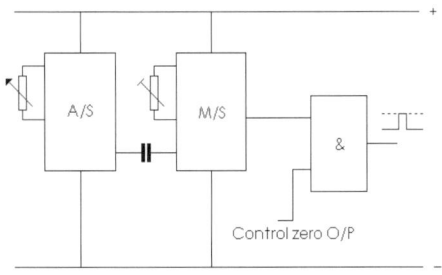

Fig.54 shows a simple PTM system in block form.
The astable triggers a retriggerable monostable via a capacitor. When the astable time period is less than the monostable time period the retriggering will cause the output to remain on permanently. When the astable time period is greater than the monostable time period the triggering will cause the output to switch in a series of pulses the relative output, like PWM, will be the ratio of the on time to off time. With PTM it is usual to make the fixed pulse of a length that ensures the motor will start turning. Depending on the loading on the motor, long variable time periods can give the effect of a step or 'jog'. This feature can be useful in systems using a DC motor for positioning.

'Smooth' PWM and PTM

With PWM and PTM the output is in many ways similar to an AC voltage when inductances are involved. It is sometimes desirable that the PWM and PTM output is translated to a discrete smoothed DC level. This can be achieved by feeding the output to a suitable size capacitor for the load. This is similar to smoothing a rectified AC power supply.

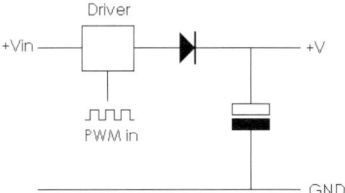

Fig.55 shows the layout for using a PWM or PTM input to produce a DC voltage relative to the PWM or PTM on/ off ratio.

Output drivers

Many power components can be used for the outputs e.g. power transistors, Darlington power transistors and MOSFETs.
The device I normally use is the HUF75337P3 N channel UltraFet. This has an on resistance of 0.014 ohms and is rated at 62A at 55V. All this is contained in a TO220 package.
MOSFETs are easy to use because they can be run in parallel to give increased current output. The one problem with the HUF75337P3 is the heatsink tab. This is not isolated and it is necessary to use an insulating washer when mounting multiple components on a heatsink.
Fig.56 shows the simplest form of MOSFET driver. This uses a diode to dissipate the

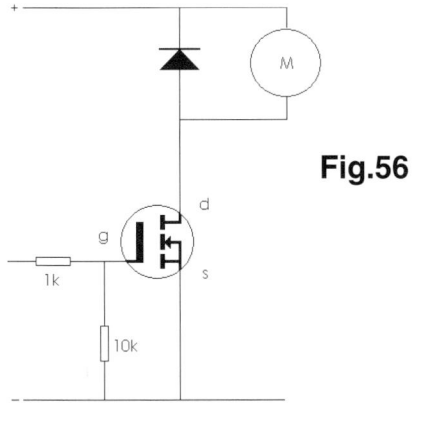

Fig.56

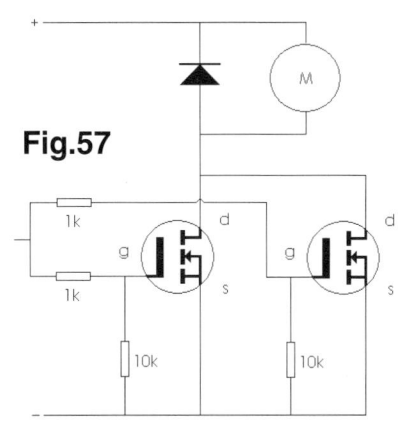

Fig.57

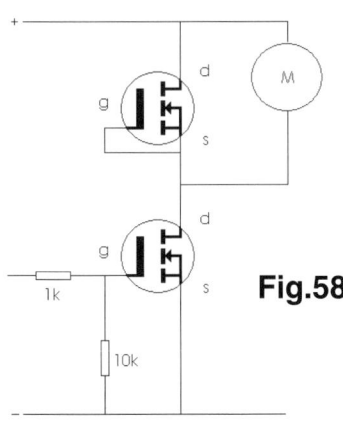

Fig.58

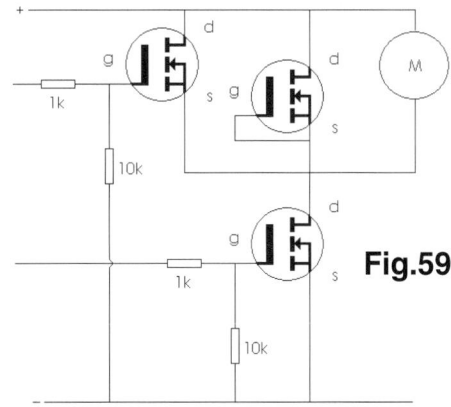

Fig.59

inductive voltage from the motor each time it switches off. The diode must be capable of carrying the current from the motor.

Fig.57 shows the method of using parallel MOSFETS to increase the current carrying capacity.

Fig.58 shows how a second MOSFET can be used instead of a diode. The MOSFET in this application works better than a diode because of its speed and conduction characteristics.

Fig.59 shows the addition of another MOSFET to provide braking.

With the MOSFET used across the motor to provide braking care must be taken in the design of the circuit to prevent this switching on at the same time as the main drive MOSFET is switched on otherwise a short circuit will occur. The braking MOSFET short circuits the terminals of the motor with the effect that the back EMF produced by the motor is fed back to the motor at opposite polarity to the direction it was running. The effect is the motor comes to a

standstill very quickly.
Output driver with brake and MOSFET 'freewheeling diode'

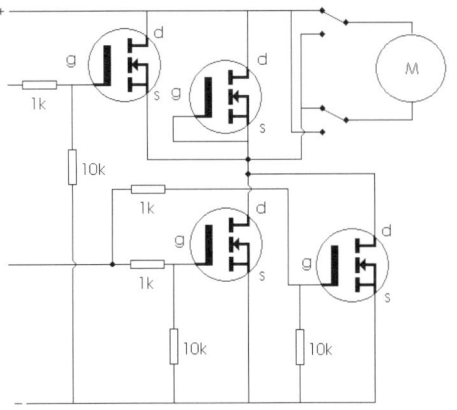

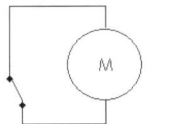 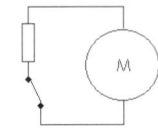

Fig.60 shows an output with two power MOSFETS used for the PWM control of a motor with another MOSFET with its gate connected to source to act as a freewheeling diode. A fourth MOSFET is used for braking but this may be optional depending on the application. The relay coil and coil driver is not shown because these are part of the voltage and direction selection circuit.

Care must be taken with this and the previous motor drive circuits to prevent switching direction when the motor is running. This can cause problems both mechanically and with the production of high reverse voltages that can damage driver components depending on the types used.

A simple joystick type or interlock control can be easily made that ensures that the speed PWM voltage is zero and braking occurs before the direction control switches over.

Hard and soft braking

It is not always desirable that the motor stops so quickly especially if it is used in a hand held device. A powerful motor stopping suddenly from a fast speed can cause a wrench on the hand holding it.

Fig.61 shows the effect of using a short circuit across the motor. The back EMF is dissipated immediately. Adding a resistor in the braking circuit means the back EMF will dissipate slower. The effect is smoother braking. The choice of resistor is mainly trial and error but a good starting point is 1 ohm. The larger the resistor used the longer will be the stopping time. The power rating of the resistor needs to be large enough to carry the power dissipated. This will depend on a number of factors e.g. the switching cycle of the motor. When used on an intermittently switched motor a small power resistor may have time to cool down between switching cycles. The same resistor used on a fast switching cycle may 'burn out' with total loss of braking. From experience I have found that 5 to 10 watts rating is a good starting point. The MOSFET braking circuit shown in **Fig.60** can be switched on to provide hard braking or a resistor can be fitted in the in the drain connection to provide soft braking. Alternatively the gate can be provided with a chopped input to provide variable braking. The time frame and duty cycle of the input will determine the level of braking. This type of dynamic braking will only work when the motor is rotating and producing a back EMF.

Motor direction control

It is easy to control the direction of a DC motor using either relays or an H-bridge configuration. With relays one relay can be used to switch power to the motor and a second double pole double throw relay can be used to control direction. Two single pole double throw relays in opposed set up can be used to control both switch on and direction. This second method has the advantage that only one coil is activated at

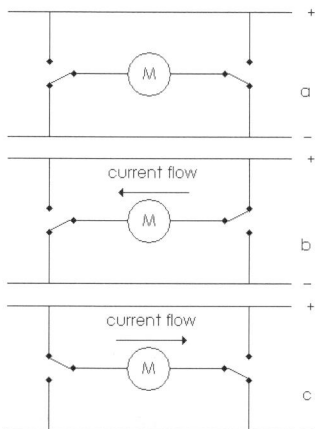

Fig.62

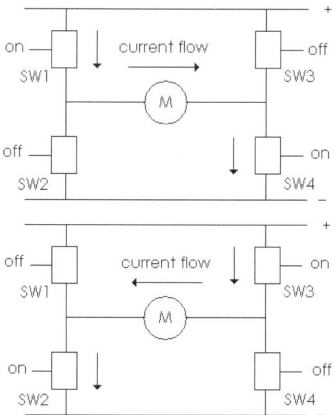

Fig.63

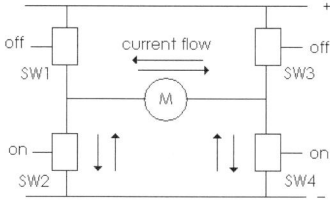

Fig.64

one time. The second advantage is that in rest position the terminals of the motor are shorted via the ground line and this can be used to effect motor braking. Accidental activation of both relay coils will have no effect because they will both move to positive potential.

Relays are rated on their current switching capability. This is the resistive load and the inductive load is a fraction of the resistive load. Typically, it may be only one tenth. Relay contacts can produce arcing and a large relay may typically have a closing time of 20mSec and an opening time of 5mSec. These times must be taken into account when compared with the circuit operating speed. It is possible to set the circuit such that zero current is switched when the relay changes state. This is easily achieved by setting the PWM to zero during relay switching operations and only switching the PWM to the set level when the relay has changed state. In this case it is probably feasible to use the relay at the resistive load rating or above.

An electronic H-bridge configuration can be either a dedicated IC or can be built from discreet components such as Darlington transistors or MOSFETs. Care must be taken with the design of H-bridge circuits to ensure that it is not possible for opposite sides of the bridge to be on at the same time otherwise this will cause a short circuit across the power lines. Free wheeling diodes are also required to reduce the inductive spikes.

As with PWM, the microprocessor can be used to provide all the necessary drive and interlock functions much easier than by using conventional circuitry.

It is also easy to build extra features such as temperature sensing into the microprocessor to monitor the temperature of the motor.

Motor direction control using a relay circuit

Fig.62 shows how two relays can be used to provide direction switching. Hard braking will be automatic when the activated relay is released. There is obviously a finite time that the relay takes to release and this must be taken into account with the speed of the motor.

Motor control using an 'H' bridge circuit

Fig.63 shows the layout for using an 'H' bridge for DC motor direction switching. When discrete components are used for the control of the 'H' bridge standard logic ICs such as inverters and exclusive or gates can be used for the required switching and braking options.

Fig.64 shows the effect of using a braking system with an 'H' bridge for DC motor. The current flow will depend on the direction the motor was turning prior to being switched off but will be opposite to the direction the current was

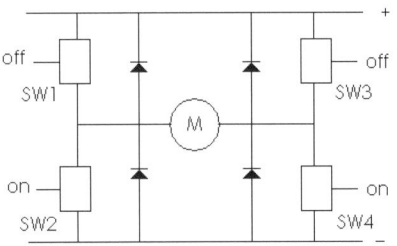

flowing and will therefore oppose the motor turning.

Fig.65 shows the motor fitted with suppression diodes. This is very similar to the bipolar stepper motor drive. Four diodes are used because current can flow in either direction through the motor. The table below shows the possible combinations for driving the switches. Of the 16 possibilities, 9 can be used the other 7 possibilities result in a short circuit with the blowing of fuses or components. Only two combinations are valid for driving. The reason that a single switch provides braking when four diodes are used for suppression is that a re-circulation path for the current is provided via the diodes and the switch that is on.

Many simple circuits use inverters to reduce the control inputs to just two instead of four. This means that a true input will switch on one pair of drivers and will switch off the other two and vice versa for the other pair. This set up works but can limit the amount of control that the above table offers. Circuits can also be produced that use the source drain junction of the switched on high-end driver and the switched off low-end driver to turn on the appropriate low-end driver. These seem to work with certain MOSFET types and motor combinations but there is the possible problem of chopping caused by the noise spikes on the motor. This source of gate drive is potentially 'noisy'.

Combining PWM and 'H' drives

There is often a need to control both the speed and direction of a DC motor. The direction can be

SW1	SW2	SW3	SW4	Result
Off	Off	Off	Off	No output or motor 'coasting'
On	Off	Off	Off	No output (braking when 4 diodes used)
Off	On	Off	Off	No output (braking when 4 diodes used)
Off	Off	On	Off	No output (braking when 4 diodes used)
Off	Off	Off	On	No output (braking when 4 diodes used)
On	On	Off	Off	'Bang' – short circuit
On	Off	On	Off	Braking
On	Off	Off	On	CW rotation
Off	On	On	Off	CCW rotation
Off	On	Off	On	Braking
Off	Off	On	On	'Bang' – short circuit
Any other combination of three or four drivers switched on will result in a short circuit with the blowing of fuses or components					

controlled by a relay and the speed by PWM or a 'H' bridge circuit can be used.

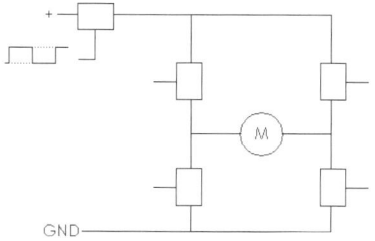

Fig.66 shows one method of using PWM with a 'H' bridge circuit. A driver is used to 'chop' the voltage and hence the power being applied to the overall motor circuit. This diagram would require the normal drive levels as with other 'H' bridge and power step circuits.

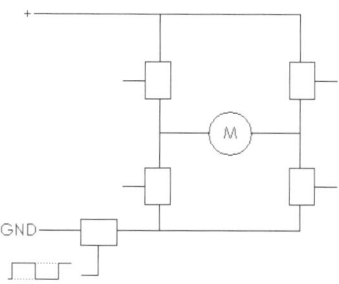

Fig.67 shows the alternative positioning of the 'chopper' circuit on the low side.

The drive codes to the 'H' bridge gates remain unchanged for either circuit. A more elegant solution is to use the 'H' bridge drivers to also control the motor speed. This also keeps the number of power components within the 'H' bridge to a minimum.

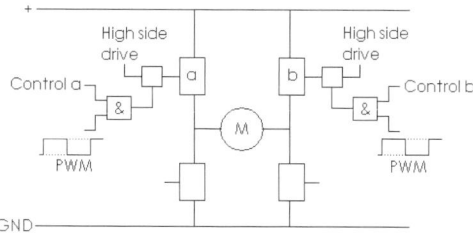

Fig.68 shows a simpler way of using existing driver components to combine PWM and 'H' bridge drive. This method uses extra 'and' gates but does not use the extra power gate with the associated drive and heat sinking. The existing high side drivers for the two high side MOSFETs, which will already be driven from a logic level voltage are driven via an and gate. The normal control sequences apply and the drive to the low side MOSFETs are unaffected. The second gate term in each of the and gates is driven from the PWM. The PWM input will have no effect unless the relevant control term is also selected. The high side drive will then cause a chopped drive via the high side MOSFET to the motor.

This type of circuit is easily controlled from a centre position joystick where the direction and speed are dependent on the direction and distance the joystick moves. Functions such as braking available by turning on the two low side drivers are unaffected.

This type of drive can also be configured in to the low side drivers.

Single phase, two phase and synchronous rectification for PWM MOSFET 'H' bridge circuits

There are a number of ways of applying drive to the 'H' bridge circuit as shown previously. The most common is to apply PWM drive to the low side driver required and turn the high side driver of the pair on fully. Or apply PWM drive to the high side driver required and turn the low side driver of the pair on fully. A less common way is to apply PWM drive to both the low side and high side driver pair at the same time.

In all these cases the stored inductive power in the motor will require a free wheel diode to dissipate the energy stored in the motor.

A neater solution is to use the existing 'H' bridge components instead of the freewheeling diodes. This leads firstly to lower cost because large current high-speed diodes are relatively expensive. There is also lower heat dissipation in the driver components because the lower resistance of the MOSFET gives a lower voltage drop.

Fig.69 shows standard low side PWM drives,

DC Motor Drive 49

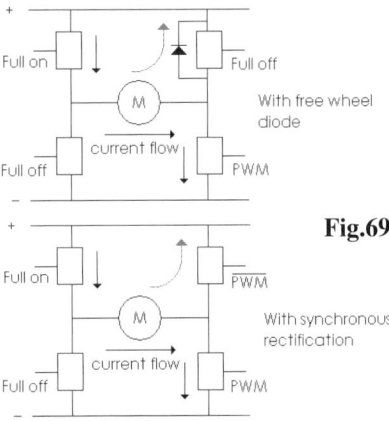

Fig.69

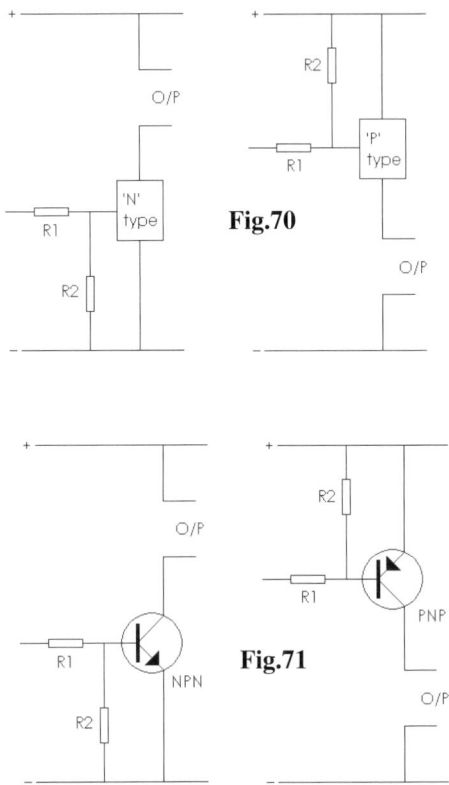

Fig.70

Fig.71

firstly with free wheel diode and secondly with active synchronous rectification.

In the first case when the PWM switches off, the energy in the motor is dissipated through the diode with the consequent high heat dissipation due to the relatively high resistance of the diode. With active synchronous rectification the high side driver above the low side PWM drive is turned on when the PWM is turned off. Delays must be built in to allow for the switching time of the components so that no overlap occurs causing both of these drivers to be switched on together causing a short circuit.

This control is easily set up using inverters and gates to give inversion and a propagation delay. There are proprietary ICs such as the TD340 with this feature built in or it is easily written into a microprocessor program.

Driving from different voltage power rails
Most electronic components switch on when a voltage is applied to the control input relative to the relevant power rail.

Fig.70 shows the comparison on 'P' and 'N' type MOSFET devices.

N channel MOSFET

With the N channel MOSFET switching is with a positive voltage on the gate relative to the source. The source is usually tied to the ground rail and therefore is a positive voltage relative to ground. The gate voltage varies between component types and is usually specified as a plus / minus voltage and a threshold voltage.

P channel MOSFET

With the P channel MOSFET switching is with a negative voltage on the gate relative to the source. The source is usually tied to the positive rail and therefore is a negative voltage relative to the positive rail. The gate voltage varies between component types and is usually specified as a plus / minus voltage and a threshold voltage.

Fig. 71 shows the comparison on 'P' and 'N' type transistor devices.

NPN transistor

The NPN transistor like the N channel MOSFET operates with a positive voltage on the base relative to the emitter. This voltage is 0.6V positive for fully turning the transistor on. Normal logic switching voltages can be used but the bipolar transistor is a current driven device unlike the FET that is voltage driven. Gain of the transistor is the ratio of the base current to the collector current. The base current resistor is chosen to limit the input and hence the output current.

PNP transistor

The PNP transistor like the P channel MOSFET operates with a negative voltage on the base relative to the emitter. This voltage is 0.6V negative for fully turning the transistor on. Normal logic switching voltages can be used but the bipolar transistor is a current driven device unlike the FET that is voltage driven. Gain of the transistor is the ratio of the base current to the collector current. The base current resistor is chosen to limit the input and hence the output current.

Power transistors compared to power MOSFETs

The power MOSFET is usually rated in terms of current capability, dissipation and on resistance. The power transistor is usually rated in terms of current capability, dissipation and collector to emitter voltage drop.

The voltage drop on a power transistor can be significant and is typically in the region of four volts. This is very large compared to the voltage drop across a MOSFET in a typical driver circuit. Because of the low on resistance and hence low dissipation and the ease of use of voltage as opposed to current drive inputs the power MOSFET is replacing the bipolar power transistor in many power circuits.

The power MOSFET because of its current sharing characteristic is easily paralleled for higher current carrying capability. Because of the low on resistance it is the obvious choice for 'H' bridge circuits. A pair of power transistors with a collector to emitter voltage of four volts each would have a significant effect on driving a motor from low voltage power rails.

The bipolar transistor is a current driven device and the current output is the current input to the base multiplied by the gain. This output is mainly linear until saturation is reached. In the linear region the heat dissipation is a factor of the effective 'resistance' of the transistor.

With a MOSFET there is a constant resistance region and a constant current region. Between these two regions is a slope that is mainly linear. On the data sheets of MOSFETs a specification called V_{GS} (threshold) is shown. This indicates the start of the switch on but the saturated state may be typically 3 to 9 volts above this dependant on the MOSFET type. This threshold voltage varies between types and may be less than 2V in small power types to greater than 6V in other types. This information is usually in the form of minimum, maximum and typical threshold voltage. The fully on gate voltage also varies between types. The maximum gate voltage is typically shown as either +/- 15V or +/- 20V depending on type although other values do exist. The gate has capacitance that it is necessary to charge at switch on and discharge at switch off. Zener diodes are often used on the gate to prevent excessive voltage pulses damaging the gate. On some ranges of MOSFET this protection may be built in to the component. It is usually positive voltages that are at risk of exceeding the maximum gate voltage but where the negative voltage on the gate is at risk of exceeding the maximum gate a second zener can be fitted in reverse to the positive level protection zener. It is wise always to consult the data sheet when using a new component type. These usually contain graphs showing the relationship between threshold voltage and saturation voltage.

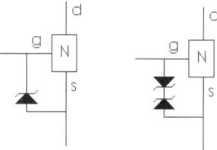

Fig. 72 shows a MOSFET with zener diode

protection with firstly over plus protection and secondly with over plus and over minus protection. In theory the zener diode could be rated at 20V (for a 20V type) but because the 20V is the absolute maximum it is probably better to err on the side of safety and choose a zener voltage a little below the 20V.

Some MOSFETs also have 'freewheeling' diodes built in. This will save the cost of a number of relatively expensive components and make circuit construction easier.

'H' bridge circuits

Unfortunately things are never as simple as they may first appear.

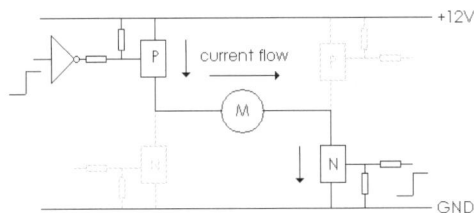

Consider the above circuit **Fig 73**. The drivers are made from complimentary pairs i.e. a matched P and N type MOSFET. This is designed to have positive going inputs to switch on. This is normal for the N channel MOSFET but is achieved by the use of an inverter for the P channel MOSFET. Positive logic is usually safer than negative going logic because the drivers need to be switched on when required, not held off until required on. With negative logic any failure of the drive circuit results in all the drivers being switched on together with the obvious results. This can be important with some microprocessor circuits because they often have a short switch on delay after power up to prevent false outputs being generated during the voltage rising to the correct logic level.

The situation is complicated by the fact that the range of complimentary paired power MOSFETs is very limited and restricts circuit design

therefore many circuits are designed using only N channel devices. Circuits could be designed using only P channel devices but there is generally a wider range of N channel devices available.

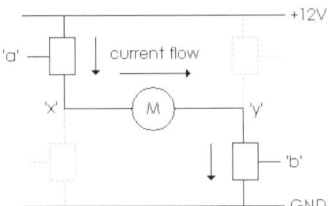

Consider the above circuit **Fig.74**. The gate resistors have been omitted for clarity. This diagram is made solely from N channel MOSFETs. The explanation is slightly simplified but is correct in practical terms. Assumptions have been made that the switching devices have no on resistance for ease of understanding. When +12V is applied to input 'b' of the MOSFET it is positive compared to the source attached to ground and the MOSFET switches on. When +12V is applied to input 'a' of the MOSFET it is positive compared to the source attached to the motor at point 'x' and the MOSFET will turn on. The motor will then run. If instead of 12V the 'H' bridge were driven from a five-volt logic source the following would occur.

When +5V is applied to input 'b' of the MOSFET it is positive compared to the source attached to ground and the MOSFET switches on. When +5V is applied to input 'a' of the MOSFET it is positive compared to the source attached to the motor at point 'x'. Point 'x' would at this time measure zero volts because no current is flowing. The gate at point 'a' is positive compared to the source at point 'x' and the MOSFET would turn on. Current would flow and depending on the motor characteristics a voltage appears across the motor and a back EMF is also produced. Assume for this example that the voltage across the motor is 8V. Point 'x' will now be 8V positive. The gate at point 'a' is 5V positive therefore the gate is 3V negative relative to the source at point 'x' and the MOSFET will turn off. In practical terms because the switching is so fast nothing will appear to happen. If point 'x' is measured with a meter or an oscilloscope then a small voltage is present. The MOSFET is acting in a similar way to a series pass transistor in a power supply.

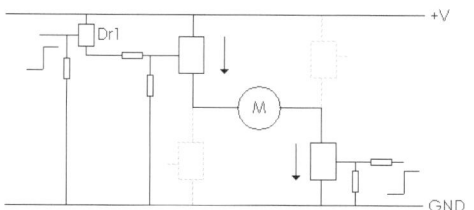

Fig.75 shows a driver made from an electronic switch such as the CMOS 4066. When a logic voltage is applied to the control term of the switch the gate of the MOSFET is connected to the positive rail via its gate resistor. The voltage on the gate will always be more positive than the voltage on the source and therefore the MOSFET will be turned on. This additional circuit would need to be duplicated on the other drivers connected to the positive rail.

The problem with this diagram is the maximum voltage that can be applied to a CMOS IC is in the order of 15V. Therefore the practical use is limited and precludes voltage switching with bipolar stepper motors.

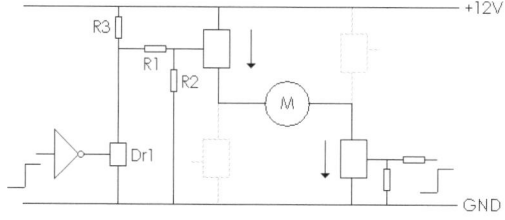

Fig.76 shows a slightly different and practical approach for using an N channel power MOSFET driver.

R1 and R2 are the normal gate resistors for the power MOSFET. R3 is a resistor that holds the gate at positive and therefore switched on. The gate will always be more positive then the source irrespective of the voltage on the power rail. Dr1 is a small low power driver capable of withstanding the maximum voltage used. It could be an NPN transistor but the preferred option is a small MOSFET. A typical example is the 2N7000 this is in a TO92 package and is rated at 60V drain to source voltage, 20mA current and a 5W on resistance. This like the power MOSFET is a voltage-controlled device and is driven via a logic level inverter in this diagram. Because the source or Dr1 is attached to ground a low voltage drive is all that is required. Gate resistors are not shown in the diagram for the small MOSFET but are a wise choice. When the input of the inverter is taken positive the output goes negative turning off Dr1. This allows the full rail voltage to appear on the power MOSFET gate turning it on. When the input of the inverter is taken negative the output goes positive turning on Dr1. This pulls the power MOSFET gate to ground turning it off. Suitable value for R3 is between about 47KW and 100KW. This additional circuit would need to be duplicated on the other drivers connected to the positive rail. It should be noted that the gate of the high side drivers are shown as being tied to ground via a resistor. This is not always practical depending on the type of MOSFET used and the current capability of the high side drive. Negative switching spikes can be put onto the gates of the MOSFETs. Using MOSFETs with back to back zener diodes either internal or external can suppress these spikes. Also the charge pumps used for the low current high side drive may have the voltage pulled down because of the current taken to ground. Higher value resistors can be used but because the switching of MOSFET in the circuit shown is the relationship of the voltage between the gate and the source, the resistor can be tied to the source instead of ground. The choice of methods will depend a great deal on the motor voltages being controlled.

Other high side driving methods

With high or variable voltage and high current drivers for digital applications it is essential that the MOSFET is used driven to saturation both for efficiency of the circuit and for reduction of heat dissipation in the driver. A gate voltage is required that is in the order of 15V above the source voltage. This voltage is chosen to try to ensure that most MOSFET drivers will work and to produce a standard circuit. Some MOSFETs reach saturation at lower voltages therefore circuits can be modified if required.

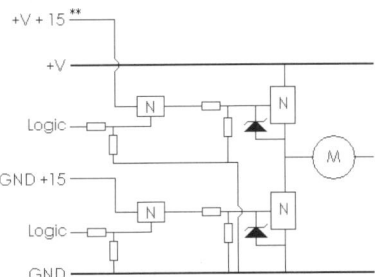

Fig.77 shows half of an 'H' bridge circuit. The thick power lines are current carrying the thin power lines are low current. The circuit shows two low current 'N' type FETs switching higher voltage on to the gates of 'N' type drivers. The sources of both low current 'N' type FETs are at ground potential via the gate leakage resistors on the power MOSFETs. This means that they will switch from simple logic drives. When either of the low current 'N' type FETs switch on a voltage will be supplied to the gate of the corresponding power MOSFET. In the case of the low side power MOSFET this will be ground plus 15V and in the case of the high side power MOSFET this will be +ve plus 15V. In many applications depending on the motor being driven and the high side power MOSFET source voltage this voltage will be too high. In most practical applications this can be reduced to +ve plus 5V.

This power supply may look complicated but each voltage level can drive a number of units. A typical low power MOSFET for high side and low side driving is the 2N7000. This is in a TO92 package and is capable of switching 60V at 0.2a with a maximum dissipation of 0.35W.

Embedded microprocessors are becoming common in everyday articles. They are found in engine and fuel management systems, many white good and even toys. It is becoming more common to be able to buy 'off the shelf' ready programmed microprocessors in the same way that you can buy more conventional ICs. Sometimes these may be in the form of kits where the printed circuit board and possibly all other components may be included.

Current limiting for motor drive circuits

Current limiting can be achieved with uncontrolled motor drive or PWM motor drive circuits in a similar way to stepper motors. If the current is monitored it is possible to change the PWM ratio or to control the V2 voltage or to cause shut down of the driver circuit. The V2 voltage can be from a multi voltage supply.

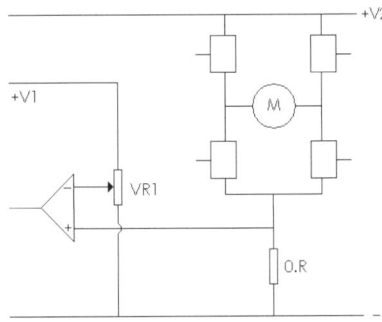

Fig.78 shows a standard MOSFET 'H' bridge driver circuit with a very low value resistor in the ground circuit. A voltage comparator is used to measure the voltage developed across the resistor and hence the current passing through it. This voltage is compared with the voltage on the threshold voltage from VR1. Changing the +/- inputs over will cause a falling instead of a rising output. As the circuit stands if used with PWM it will trigger on the peak current taken by the motor.

A method often used with peak current sensing is to use the output of the comparator to trigger a monostable that inhibits the output driver. At the end of the monostable timing cycle the output

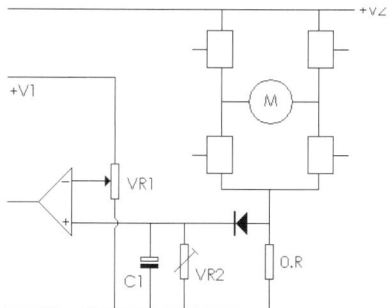

driver switches back on and the cycle is repeated each time the current limit is reached. The drive is therefore in the form of a chopped output.

This will give a form of speed control for a fixed load but as the load increases the torque will fall because the current is limited.

Fig.79 shows modifications to the previous circuit that will give triggering at a level nearer to the average current taken by the motor instead of the peak current. C1 is charged via the blocking diode. VR1 provides a leakage path for the capacitor. This is necessary because the high impedance gates on the voltage comparator would mean that the capacitor would eventually

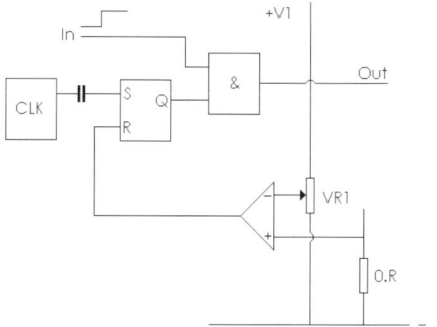

charge to the peak voltage. C1 and VR2 are best chosen by experiment to match the circuit usage. A reasonable starting point would be 1mF and 100KW.

Current chopping

Fig.80 shows a circuit for current chopping for motor circuits. It can be used with DC motors but is more common with stepper type motors. The circuit is similar to the type used in a number of stepper motor driver ICs. The circuit is shown using discrete components and logic gates. The current sensing is the same as in the previous circuits but this time it resets a set/reset latch. The latch output is an input to an and gate. The output to drive the power control circuit is via the and gate. The other input to the and gate is the 'in' control term.

When the current limit is reached the latch is reset and the output is turned off stopping the current flow. A continuously running clock circuit of about 2KHz turns the latch back on allowing current to flow. The latch is then reset again when the current limit is reached. The cycle is repetitive maintaining the current at a reasonably stable level.

The clock can be built around a simple 555 astable circuit or can be derived from a suitable clocking point elsewhere on the circuit.

On standard 2/4 coil unipolar and bipolar stepper circuits it is usual to have a resistor and sensing circuit per pair of coils. In multiple coil variable reluctance type circuits using wave drive where only one coil is on at any one time it is feasible to use only one resistor and sensor circuit. The output from this can feed a latch/and gate combination for each coil. The clock can feed a large number of latch circuits.

Because 324 OP amps, 2 input and gates and simple SR latches have four circuits per IC the only real saving is the resistor.

Prevention of motor direction switching with motor running

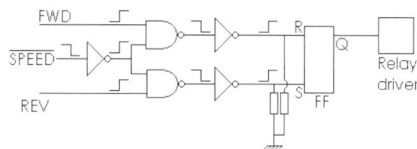

Fig.81 shows a circuit to prevent motor reversal until the motor is stationary. A positive triggered flip-flop is used to switch a relay circuit between

notional reverse and forward. Forward is assumed to be the direction when the relay is not activated. A speed or current sensor circuit as described previously provides a speed input to one gate of two nand gates. This prevents the forward or reverse inputs 'triggering' the nand gates unless the speed input is below the set threshold. Suitable 2 input nand gates are the 4011 and the inverter can be a 4049. The flip-flop is any positive triggered CMOS flip-flop or set/ reset latch or can be made up from individual gates.

Practical DC motor MOSFET drive circuits

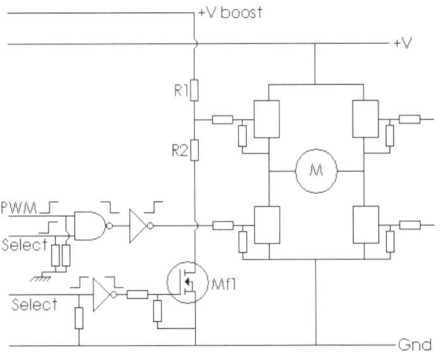

Fig.82 shows a practical MOSFET 'H' bridge drive. The high side drive is from a boost voltage. Mf1 is a small MOSFET such as the 2N7000. This is capable of switching 60V drain to source with a drain current of 200mA. It comes in a TO92 package. Mf1 is switched from an inverting buffer. With no input the inverter switches on MF1 and R2 is pulled to ground. The junction of R1 and R2 is +ve boost multiplied by the ratio of the resistors. Suitable starting values are 10K and 4.7K for R1 and R2 respectively. This provides a gate voltage of about one third +ve boost. With a positive input to the inverting buffer Mf1 is switched off and the junction of R1 and R2 rises to +ve boost. The circuit is duplicated for the right hand drivers. Suitable gates are 4011 for the 2 in nand gate and 4049 for the inverting buffer. A suitable power supply must be provided for the CMOS logic. The gate and gate to source resistors are as

described in the theoretical section on the MOSFET 'H' bridge. PWM is applied to one input of both low side drivers consecutively.

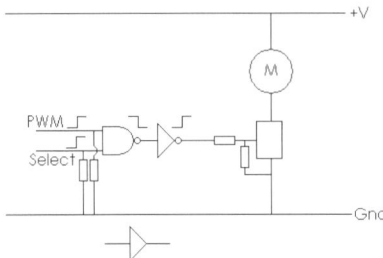

Fig.83 shows a section of a practical MOSFET drive for a DC motor where reversing is not required or is relay controlled. The small circuits shown can replace the 2 input nand gate if the PWM will provide a zero level input and 'select' is not required. The non-inverting buffer is a 4050.

A practical unit for use with multiple motors

The following is a description of a simple unit that can be combined with a driver to give a unit that automatically selects the motor voltage and the PWM minimum and maximum settings when the motor is plugged in. It is particularly useful when using a number of low voltage power tools. It can be used with the latching or 'first on' switching described in the switch section and a separate PWM controller from the motor speed

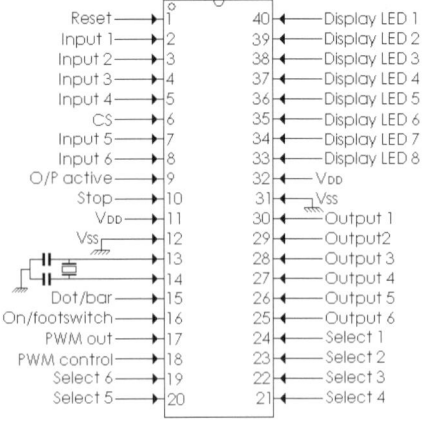

Fig.84

\multicolumn{3}{c}{Pin designation}		
Pin	Pin name	Description
1	Reset	Resets when grounded, tie to + rail if not used.
2	Input 1	Analogue input 1.
3	Input 2	Analogue input 2.
4	Input 3	Analogue input 3.
5	Input 4	Analogue input 4.
6	CS	Chip select, active positive, tie to + rail if not used.
7	Input 5	Analogue input 5.
8	Input 6	Analogue input 6.
9	O/P active	Turns on in conjunction with PWM control. Can drive low power LED or LED driver.
10	Stop	Sets all outputs to off. Can be used as an emergency stop. Tie to + rail via 10K resistor, pull to - rail to activate, or tie to + rail if not used.
11	V_{DD}	Positive rail connection *[1]
12	V_{SS}	Zero or ground connection
13	OSC 1	Oscillator input 20MHz.
14	OSC 2	Oscillator input.
15	Dot/bar	Selects moving dot or moving bar output for LED display.
16	On/footswitch	Activates PWM control. Tie to - rail via 10K resistor, pull to + rail to activate, or tie to + rail if not used.
17	PWM out	Pulse width modulation output.
18	PWM control	Gating control for use with PWM output. Works in conjunction with select and on. Allows PWM minimum to be above zero.
19	Select 6	Momentary on latches matching output.
20	Select 5	Momentary on latches matching output.
21	Select 4	Momentary on latches matching output.
22	Select 3	Momentary on latches matching output.
23	Select 2	Momentary on latches matching output.
24	Select 1	Momentary on latches matching output.
25	Output 6	Output to power relay driver.
26	Output 5	Output to power relay driver.
27	Output 4	Output to power relay driver.
28	Output 3	Output to power relay driver.
29	Output 2	Output to power relay driver.
30	Output 1	Output to power relay driver.
31	V_{SS}	Zero or ground connection.
32	V_{DD}	Positive rail connection. *[1]
33	Display LED 8	Output to LED driver for dot/bar display.
34	Display LED 7	Output to LED driver for dot/bar display.
35	Display LED 6	Output to LED driver for dot/bar display.
36	Display LED 5	Output to LED driver for dot/bar display.
37	Display LED 4	Output to LED driver for dot/bar display.
38	Display LED 3	Output to LED driver for dot/bar display.
39	Display LED 2	Output to LED driver for dot/bar display.
40	Display LED 1	Output to LED driver for dot/bar display.

*[1] *The positive voltage range is +3V to +5.5V, but it is recommended that a regulated +5V power source is used. Inaccuracies will occur if variations between the IC supply and the input supply occur.*

Maximum analogue input is V_{DD}. Attenuation must be used with inputs exceeding this value. Over voltage input protection is recommended because the device may be damaged with voltage exceeding V_{DD}.

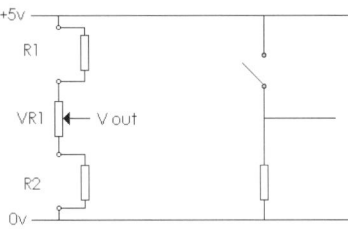

Fig.85

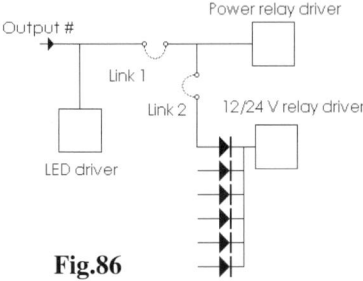

Fig.86

control section. It can also be used with the dedicated microprocessor unit described below. If used with the discrete devices it is necessary to use power relays for the power and a small reed or similar low power relay to switch the input to the PWM driver.

The BM201P 6 input 6 output PWM controller **Fig.84** shows the layout of the BM201P

General description

The BM201P is intended for the driving of up to 6 small DC power tools from a single power supply.

The IC is a dedicated PWM 6 input and 6 output controller. It produces a latched output for a momentary input on the matching select pin. The outputs can be used to switch power relays via a suitable relay driver to feed current to a specific output.

Gating control allows switching of PWM output with external gating. This feature works in conjunction with on/footswitch to allow PWM minimum to be above zero when starting.

There are six individual analogue inputs and only one output can be latched at a time even if more than one select button is pressed. The lowest number select pressed will activate.

Output level indication for an eight-bit LED display using appropriate driver. The output gives an 8 bit output relative the 100% PWM range. There is a dot/bar selection for the display.

Dual voltage output can be selected automatically to match requirements of the motor connected.

There is a 'stop' pin input. This can be used with inputs such as emergency stop or can be used for options such as over temperature.

he PWM gating control output and the PWM output can be fed to a 2 input and gate. The PWM output from the and gate will only be true when both the PWM gating control output and the PWM output are true. Other logic devices or a MOSFET gate can achieve the same gating.

Fig.85 shows the method of setting a minimum and maximum PWM range using resistors. The resistor values can be calculated or obtained by using a test plug with R 1 and R 2 replaced with variable resistors and measuring the resistors when the required range is achieved.

If the resistors are mounted in the plug of the tool and the potentiometers are part of the power unit then the settings will transfer irrespective of the plug connection. There will be 1 potentiometer and 1 socket for each of the channels used to a maximum of 6.

Also shown is the on/ footswitch connection.

Fig.86 shows a possible output block diagram. The LED driver and the Power relay driver are duplicated for each of the 6 channels. Link 1 and Link 2 are contained within the tool plug. Link 1 acts as a 'motor present' connection, this prevents activation of the power relay and the 12/24V relay if no tool is plugged in. Link 2 is the 12 or 24V relay driver. The voltages can of course be other appropriate low voltages for the tools being used. The normal configuration is to use relay drive for the lesser used voltage. Individual relay drivers can be used but using the diode gating may reduce component count.

Power supply

Because the controller/ driver can be used in a number of applications a dual voltage power supply is shown. This makes it possible to run a range of motors from one controller. Part of the

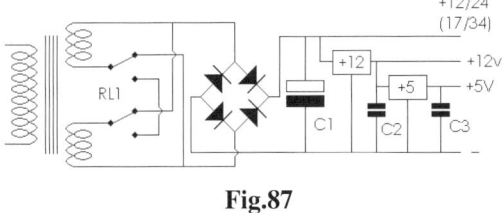

Fig.87

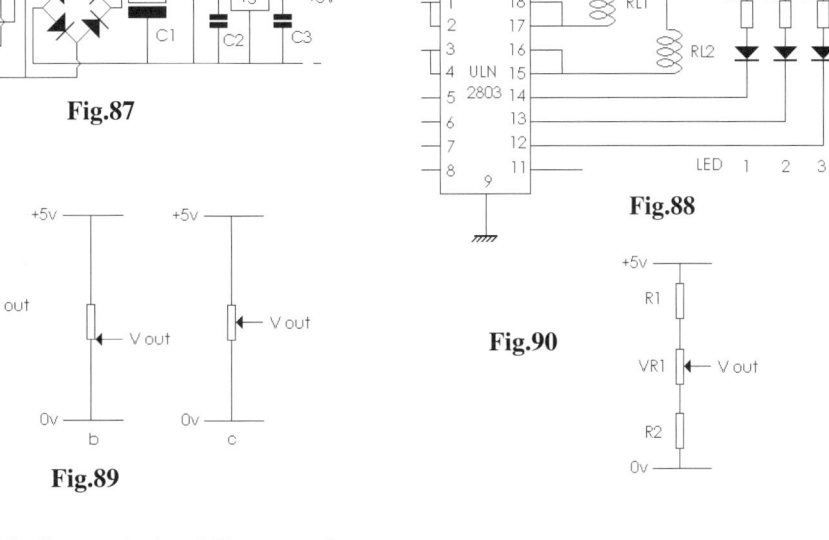

Fig.88

Fig.90

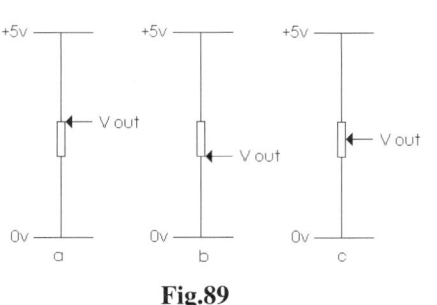

Fig.89

uniqueness of this diagram is the ability to use it with different motor units. The plugs for the motor units contain an ID which sets the maximum speed, the minimum speed and the 12V/24V selection. If the motor and controller are intended for use in a dedicated unit the parts of the circuit not required can be left out. If only the 12V power is needed for use with the model motors described earlier the 12V/ 24V selection can be left out and the transformer 'hard wired' to 12V. This method of selection is chosen to make the use of a smaller transformer possible. For example with two separate 12V 10A windings in parallel this will provide 12V at 20A. In series they will provide 24V at 10A. This is a transformer of 240VA. If a 12-0-12 centre tapped transformer were to be used then each part of the winding would need to be capable of providing 20A even though this would not be needed when 24V is selected. This is the equivalent of a transformer of 480VA. This is physically larger and much more expensive.

Fig.87 shows the power supply the +12/24V (17/34 after smoothing) is the main power for the motor. The +12V is from a 2A regulator and provides voltage for the relays and LEDs. The +5V is from a 1A regulator and tapped from the +12V and is effectively double regulated. This is the supply for the microprocessor.

Voltage and direction selection circuit

This part of the circuit contains a ULN2803 driver for relays and LEDs. If only a basic circuit were required this could be dispensed with and replaced with discrete transistors or FETs. The ULN2803 has eight separate Darlington drivers in one eighteen pin package. Each driver can sink currents of 500mA at up to 50v on a 23% duty cycle at 25°C. The drivers can be paralleled to provide up to 4A at the same duty cycle and temperature. The IC also contains internal diodes for inductive loads and base resistors therefore saving on the overall component count.

Fig.88 shows the relay circuit for the 12/ 24V select and for the direction selection circuit. The LEDs are used to indicate power on, forward and reverse. The concept of forward and reverse is purely notional because the direction of rotation will depend upon the application and the direction can be set as the desired forward by changing over the motor wires.

Automatic selection

The following specifications are set up automatically when a tool is plugged into the universal unit. On dedicated units these are built in directly.
  Minimum motor PWM %
  Maximum motor PWM %
  12/ 24V selection
  Brake/ no brake selection
  Maximum motor operating temperature – optional

The microprocessor has the ability on its analogue to digital controller input to work with reference voltages or from 'rail to rail' voltages. Using the second option means that variations in the power supply voltage to the microprocessor, provided it does not go outside the operating voltage, will not have a noticeable effect on the PWM because the values read are relative in percentage terms to the rail voltages. **Fig.89** shows analogue input relative to the PWM output.

Consider V out in the diagrams a, b and c. V out in a will be 5V because the slider is at the top of its travel. V out in b will be 0V because the slider is at the bottom of its travel. V out in c will be 2.5V because the slider is at the middle of its travel. This equates to a PWM range of 0% to 100%.

Fig.90 shows analogue input range modification and therefore PWM range modification.

If extra resistors are added, the voltage range can be modified. E.g. if R1 = R2 = VR1 the junction of R2 and VR1 will be 5 x $\frac{1}{3}$ i.e. 1.66V. Similarly the junction of R1 and VR1 will be 5 x $\frac{2}{3}$, i.e. 3.33V. This equates to a PWM range of 33.33% to 66.66%. Resistors can therefore be used to give a wide range of PWM values.

Fig.91 shows plug connection for automatic function selection.

There are a wide range of plugs and sockets available. The most widely available economical plug and socket combination I have found is the solder pin 15 way 'D' type. These are rated at 300V RMS and a working current of 7.5A per pin. Wide ranges of shells are available to fit these connectors. If more current capacity is needed,

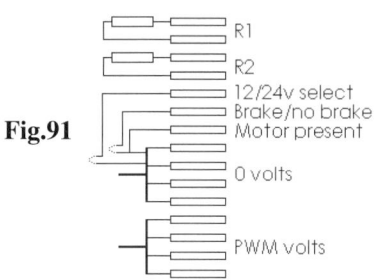

Fig.91

the 25 way 'D' type can be used. The stylised drawing shows the internal connections on the motor lead. The 12/ 24V select and the brake/ no brake options are selected by linking the pin to ground if the feature is required. The motor present is a safety precaution that prevents the PWM operating until a connection is made. If the temperature sensing option is used it is necessary to sacrifice one 0V and one PWM volts pin to allow connection. This could also be achieved by using a 25 way 'D' type. The temperature sensor is a 3-pin solid state device, one of the pins being connected to 0V. In a dedicated unit all the connections could be made in the control unit and the plug only being used for motor present, power and the optional temperature sensing.

The TD340

The TD340 is an IC that was introduced a few years ago. It is intended to provide PWM speed control and direction control with MOSFET 'H' bridge drivers. It is easily interfaced with a microprocessor or with manual controls. The TD340 is primarily intended for automobile battery voltage applications. Its voltage range is 6.5V and 18.5V but it can withstand a maximum voltage of 60V for one second. This IC contains a number of very useful features. The TD340 contains two independent charge pumps to provide the higher voltage drive for the high side driver MOSFETs. It also allows active synchronous rectification for the free wheel current. This feature uses 'dead times' and close tolerance timing to switch the high side MOSFETs to allow dissipation of the free wheel current. This means that the circuit does not need the free wheel diodes. This can be a cost saving with high current circuits but probably more important lowers the heat dissipation considerably. Brake mode is also built in to the IC. Brake mode is initiated when the input to pin 7 is at a zero level for more than 200μS. Zero level is a voltage between 0V and 1.2V. High side or low side braking is automatically chosen from the input of pin 9, the direction control. The TD340 also contains a 5V regulator, a reset circuit and a watchdog circuit for use with a microcontroller.

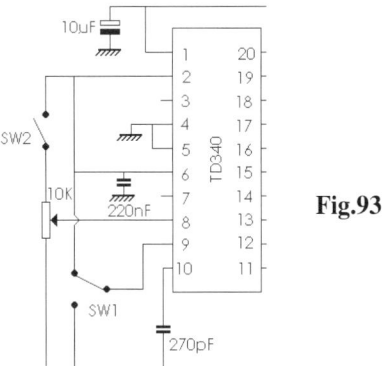

Fig.93

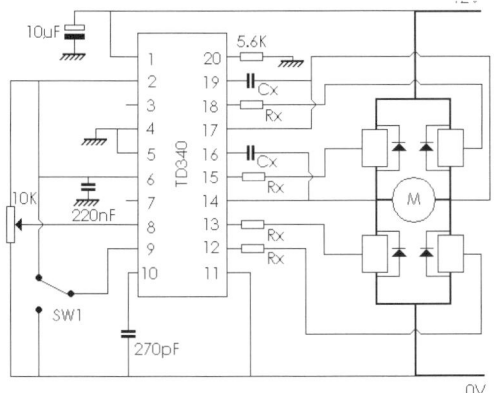

Fig.92 shows a typical circuit for manual control from a 12V power supply. The darker lines indicate the high current section of the circuit. Cx are the charge pump capacitors and are typically 22nF. Rx are the MOSFET gate resistors and are chosen to match the MOSFETs but are typically between 10Ω and 100Ω. SW1 controls the direction of drive and the 10K potential divider on pin 8 controls the PWM width. Extra resistors in the potential divider circuit can provide offsets such that the full range of the potential divider can be used.

This type of circuit is becoming more common in automobile applications for manual functions such as electric windows and sunroofs.

Fig.93 shows a very simple modification to the 'front end' of the TD340. SW2 allows the speed to be left in a set position. When SW2 is opened pin 8 will be connected only to ground via the lower part resistance of the potential divider

hence braking will occur. Care must be taken with this and the previous TD340 circuit to prevent switching direction when the motor is running. This can cause problems both mechanically and with the production of high reverse voltages that can damage driver components depending on the types used.

A simple joystick type control can be made that ensures that the speed voltage is zero and braking occurs before the direction control switches over.

Feedback speed control with the TD340

Speed feedback from a tacho can be used with the TD340. This would need a circuit that compared the required input speed setting with the actual output speed and made adjustments to the actual input speed voltage setting depending on the difference between the two voltages. This type of circuit would require relatively complex discreet electronics. It is therefore probably better to use a microprocessor for the comparisons. Interfacing between the TD340 and the microprocessor would also be needed. This raises the question of why not just use a microprocessor for all the functions. The 'special' features of the TD340 are easily built into a programmable microprocessor with the exception of the charge pumps. The charge pumps are easily made from the simple circuits described in the relevant section of the book. If a number of TD340s were being controlled

Pin designation

Pin	Pin name	Description
1	Reset	Resets when grounded, tie to + rail if not used.
2	Speed in	Analogue input to max of Vcc rail. PWM output control.
3	Boost set	Analogue input to max of Vcc rail used in conjunction with boost input to allow a PWM boost.
4	Brake	Sets 'H' bridge outputs to dynamic braking i.e. ground shorted. Switches off relay outputs. Tie to + rail via 10K resistor, pull to − rail to activate, or tie to + rail if not used.
5	Not used	Connect to − rail.
6	FWD	Notional forward.
7	REV	Notional reverse.
8	V_{SS}	Zero or ground connection.
9	OSC 1	Oscillator input 20MHz.
10	OSC 2	Oscillator input.
11	PWM control	Gating control for use with PWM output. Works in conjunction with FWD and REV. Allows PWM minimum to be above zero.
12	PWM out	Pulse width modulation output.
13	Boost	Positive input selects boost setting instead of speed in setting. Can be used when starting heavily loaded motors.
14	Brake/no brake	Selection for brake or no brake when coasting using 'H' bridge drivers.
15	CS	Chip select, active positive, tie to + rail if not used. If deselected then output will remain as is until reselected.
16	Stop	Sets all motor outputs to zero. Can be used as an emergency stop. Tie to + rail via 10K resistor, pull to − rail to activate, or tie to + rail if not used.
17	REV block	An input on this pin prevents the motor being reversed but still allows brake and coast features. Prevents motors being rapidly turned to opposite direction. Tie to − rail via 10K resistor, pull to + rail to activate, or tie to − rail if not used.
18	Brake O/P	Gives output when brake input selected. This can be used to give LED indication or be used to activate other mechanical braking.
19	V_{DD}	Positive rail connection. *[1]
20	V_{SS}	Zero or ground connection.
21	REV LED	Output that can be used to drive LED to indicate REV selected.
22	FWD LED	Output that can be used to drive LED to indicate FWD selected.
23	RL1	Relay control 1.
26	RL2	Relay control 2.
25	'H' bridge 1	'H' bridge driver 1.
26	'H' bridge 2	'H' bridge driver 2.
27	'H' bridge 3	'H' bridge driver 3.
28	'H' bridge 4	'H' bridge driver 4.

*[1] The positive voltage range is +3V to +5.5V, but it is recommended that a regulated +5V power source is used. Inaccuracies will occur if variations between the IC supply and the sensor or input supply occur.
Maximum analogue input is V_{DD}. Attenuation must be used with inputs exceeding this value. Over voltage input protection is recommended because the device may be damaged with voltage exceeding V_{DD}.
The oscillator consists of a 20MHz crystal with capacitors as recommended by crystal manufacturer. Typically 32pF.

from for example the parallel port of a PC then the exercise may be worthwhile. Otherwise it may be better to accept the TD340 as an excellent device for manual control of a motor.

The BM202P PWM controller

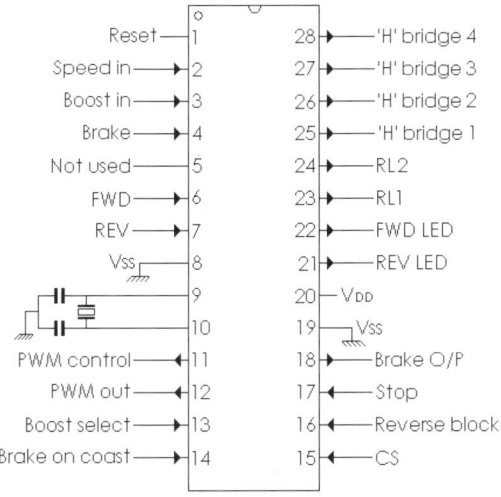

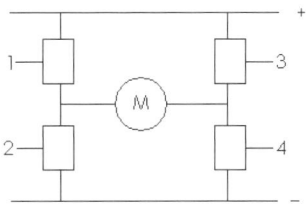

Fig.95 shows the 'H' bridge driver connections to a suitable transistor or MOSFET driver.
Fig.96 shows the relay driver connections to a suitable transistor or MOSFET relay driver.

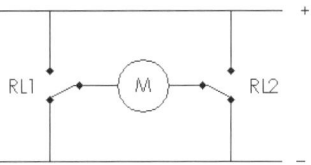

Fig.94 shows the layout of the BM202P PWM controller.

General description

The BM202P is a PWM controller for use with 'H' bridge or 2 relay motor control circuits. The IC has a number of features that make it versatile for use in a range of applications using low voltage DC motors.

Features include PWM output control, boosted PWM output control allowing PWM minimum to be above zero when starting, 'H' bridge and relay outputs and various brake and coast features for use with 'H' bridge outputs. There is also an input to prevent the motor being rapidly turned to the opposite direction whilst running in one direction. The PWM output has gating control working in conjunction with FWD and REV. The PWM gating control output and the PWM output can be fed to a 2 input and gate. The PWM output from the and gate will only be true when both the PWM gating control output and the PWM output are true. Other logic devices or a MOSFET gate can achieve the same gating.

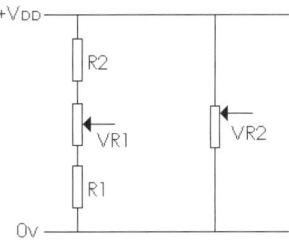

Fig.97 shows typical connections for the PWM input on VR1 and the boost input on VR2. R1 sets the lower limit of the PWM output and R2 sets the upper limit of the PWM output. The boost voltage can be set to a value up to 100% even if the 'normal' PWM maximum is set at a lower level.

Ripple DC drive

This is a form of control that may find an application in simple circuits powered from the AC mains via a step down transformer.
If the output from a transformer is rectified through a bridge rectifier but not smoothed then the output is in the form of a 100Hz ripple with a voltage going from 0V to a peak at the RMS

DC Motor Drive 63

value of the transformer output. In the case of a 12V transformer this will be in the order of 17V. A thyristor or SCR – silicon controlled rectifier – is a device that when triggered through its gate will conduct until the voltage being conducted falls to zero. It then ceases to conduct until its trigger is again taken to the required trigger voltage. The normal triggering voltage is in the order of 2V, with a trigger current that can be from fractions of milliamps to about 20mA.

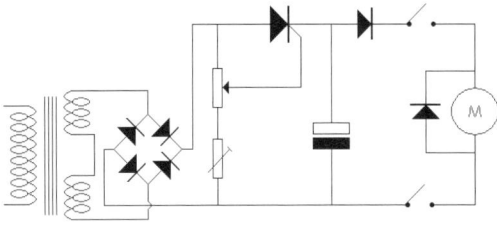

Fig.98 shows a simple form of motor control that uses a potential divider to obtain the trigger voltage. As the voltage across the potential divider rises from 0V to 17V the tapped voltage depending on the position of the potentiometer slider will reach the trigger voltage and the SCR will conduct. The trim resistor in series with the potentiometer allows full turn of the potentiometer. Without this the potentiometer would only use a small percentage of track movement to cover slow to fast. A good starting point for the potentiometer is 4.7KΩ and for the trimmer 47KΩ. These may need to be modified depending on the trigger requirements of the SCR chosen.

The rest of the components in the circuit are blocking or back EMF protection as in previous circuits. The capacitor is optional but can help to smooth the motor and prevent 'ringing' that sometimes occurs. If the SCR was used in a mains voltage circuit a diac is the normal method of triggering in this simple type circuit. The diac is a diode like device that only conducts when the voltage across it exceeds its breakdown voltage, which is in the order of 30V.

Changeover relays for direction control can be used as in other circuits but must be fitted such that the back EMF protection diode is not forward biased.

This simple circuit has a number of disadvantages. Because the triggering is from a symmetrical wave the peak occurs half way along each wave peak. Therefore the latest triggering that can occur is half way along the peak therefore the effective range is from half to full. The unit will only have a zero output if the trigger voltage is higher than trigger input voltage but noise spikes can cause spurious switching therefore in the interests of safety an on/ off switch is desirable in the output.

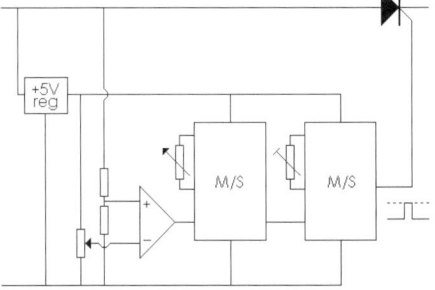

Fig.99 shows a block diagram of an improved but more complex method of triggering. This uses a voltage comparator to obtain a trigger for one monostable that produces a delay and then triggers a second monostable that produces a short pulse to trigger the SCR. The resistors on the positive gate of the comparator are chosen such that the maximum voltage from the ripple supply does not exceed the 5V supply to the comparator.

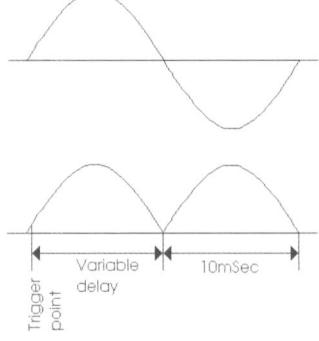

Fig.100 shows the sine wave before and after

64   Electromechanical Building Blocks

passing through the bridge rectifier. The voltage comparator is triggered on the rising part of the ripple voltage as near to zero as possible. This triggers the monostable for the delay on a rising edge. When the delay is over the output will fall and trigger the second monostable on a falling edge. This produces the trigger pulse for the SCR. This type of circuit will work but problems can occur with the first monostable delay if very low power is required. Too long a delay can actually cause the triggering to begin at the start of the next half cycle giving near full power instead of the required low power. If the maximum delay does not exceed 8mS and the initial triggering is near to the zero point this circuit will give reasonable results.

Except for very basic designs with this type of circuit it is probably better to make use of PWM driving techniques instead.

Motors as generators

The DC brush motor will act as a generator if rotated and the output is usually linear over the practical speed range. The small DC motor then provides a practical output that can be used in applications such as speed measurement to give a direct reading or for speed control in a servo speed feedback system.

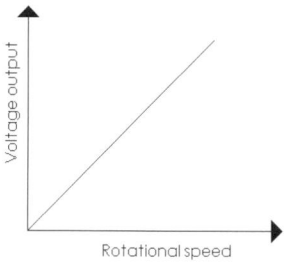

Fig.101 shows the typical output from a permanent magnet motor used as a tacho generator. The actual slope and voltage output will depend on the characteristics of the motor used. The voltage output is essentially linear but at very low speeds some non-linearity may occur.

DC Motor maintenance

Many small motors rarely need more than a small amount of light lubrication over their life. They are usually replaced when they are worn out. If the motor is of a type that replacements are expensive or difficult to obtain then work is justified.

The main problems are bearings particularly of the phosphor bronze type if the motor shaft has been subject to a side force, and in the area of the motor brushes and the commutator.

Bearings are usually a push fit and although many bush type bearings are not off the shelf sizes, they can be easily made if access to a lathe is available.

Motor brushes are easily replaced and if the size cannot be found it is easy to 'rub down' a larger brush on a piece of abrasive paper or a flat file to the required size.

The commutator is an area that has caused a great deal of argument over the years. After much use the motor brush will often wear a groove in the copper segments of the commutator. This can lead to increased brush wear and arcing. The commutator will require a skim to remove the wear. This is very simple but there is insulation between the copper segments and the argument is whether this should be flush or undercut. My stance on this is very simple, the manufacturer knows best. Look at an area of unworn commutator before starting work. If the insulation is undercut, then cut the skimmed insulation back to the same depth. If the insulation is not undercut then leave it alone.

Motor couplings

The most common form of coupling one shaft to another is by gears or belts. There is often a need to couple shafts that are in alignment end to end. The couplings then take the form of rigid or flexible. Rigid couplings are basically a section of rod with holes to match the shaft and mechanical fastenings e.g. screws or pins to fasten it to the shafts. Flexible couplings take many forms but the basic concept is a hub to

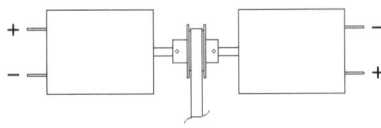

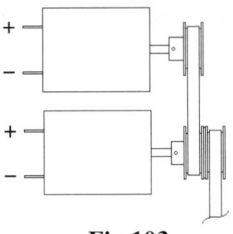

Fig.102   Fig.103

attach to each shaft and a coupling between the hubs that allows the hubs to move slightly out of alignment. The flexible section can be moulded rubber shapes, pins, flexible metal sections or bellows or section that allow movement because of right angle grooves or forks and a drive block.

Extra power from DC motor drives

Because the current taken by an individual motor is proportional to the load and the voltage used, it is possible to use multiple motor drive.
Motors can be connected by solid or flexible links with one motor running 'forward' and the other in reverse or by gears or belts. The same PWM control can be used for both motors but it is probably better to use one PWM control and individual PWM power drivers.
Provided that the second motor can be left with an open circuit it is possible to run just one motor and bring in the second motor when extra power is needed.
Fig.102 shows a pair of motors coupled by a shaft coupler with a toothed belt drive.
Fig.103 diagram shows a pair of motors coupled by a toothed belt drive. The double pulley allows drive to be taken to other parts of the machine.

CHAPTER FIVE

The Servo System

Servo is any form of system that uses a reaction to determine the outcome of an action. This may be in the form of a tacho-generator attached to a motor. The system is a closed loop where one part of the system moves to follow another part of the system. This can be either negative or positive feedback. A negative feedback system is for example a sensor attached to a motor to determine speed. When the output from the sensor increases the motor is therefore going faster so the system reduces the drive to the motor. A positive feedback system is for example a motor driving a feedback potentiometer. As the voltage increases on the position setting potentiometer the motor will drive the feedback potentiometer to increase the voltage until both voltages are equal.

Unlike a stepper motor the drive used with a servo machine system is usually continuous. A feedback system is used to determine the position of the driven component that may be some form of tool e.g. a router. The advantage of the servo system is that the positioning is directly read and is not susceptible to 'slip' or missed steps that can occur with stepper motors. The system can be used with motors that produce more power than is usually available with a stepper motor. Computer control can allow fast traverse to a position near the required final or start of operation position and then slower drive is used for exact positioning. Multi pole motors can be used to produce a range of speeds. With an AC motor the speed is a function of the poles. A 2 pole motor runs in the order of 2800 RPM on 50Hz whilst a 4 pole motor runs at 1400 RPM. Doubling the number of poles halves the speed of the motor up to a practical limit depending on the physical size of the motor. This coupled to variable frequency drive and DC braking can allow a good resolution and repeatability.

The complexity of the servo system will depend on the function of the machine. It is feasible for an automatic drilling machine to overshoot and jog back to its destination point because the action of drilling a hole only occurs when the drilling head is positioned correctly. This could not be tolerated on any machine that cuts on the move such as a milling machine, a router or a laser cutting machine. Most of the description above adds up to a high degree of complexity and large amounts of money.

Another type of servo system used in computing was the balanced coil where two opposed currents were used to set a balance point. The ratio of the currents and hence the magnetic field set a physical position for the attached linkage. This current balance system is used to achieve micro stepping in stepper motors. This balance system is also used with opposed air rams for positioning in a number of limited applications. The most common true servo system seen is likely to be in the model field. These are used for control of flying surfaces in model planes or rudders in model boats. Some are now fully digital but most use a pulse width system to determine the desired position and comparators to know when the position is reached. These non-digital systems are easy to set up in the workshop but have limitations of resolution because a 'dead area' is needed or they will continue to hunt around the position due to motor overshoot.

The digital system is more difficult to implement. A one degree absolute encoder requires nine-bit coding. This coding is usually Gray code this unlike binary uses only one bit change per step. Consider a four bit coding disc with binary coding at a transition point between 0 and 15, this produces a four bit change bit change. Even with these limited number options binary coding give a range of bit changes from one to four. With an 8 bit disc of 256 positions the 256 to 0 transition is an 8 bit change. This can produce a large number of potential errors and precautions such as centre of bit strobing may be necessary to ensure accuracy. The main advantage of the binary code is that it can be read directly by a computer or by a number of off the shelf driver and control ICs without any decoding. The Gray code is accurate because in any of the incremental changes only one bit alters so that the production of position discs is less critical as regards edge alignment. The disadvantage of the Gray code was the necessity to decode to binary for use by the processing unit. This may not have been a problem if the processing unit was a computer but smaller pieces of equipment would probably have required a decoding unit that may have been formed from a number of exclusive 'or' gates. The number of gates normally needed is bits – 1 i.e. a four bit decoder requires three gates and a nine bit decoder requires eight gates. There is also a large amount of gate

Binary value	Binary code				Gray code			
	Bit 3	Bit 2	Bit 1	Bit 0	Bit 3	Bit 2	Bit 1	Bit 0
0	0	0	0	0	0	0	0	0
1	0	0	0	1	0	0	0	1
2	0	0	1	0	0	0	1	1
3	0	0	1	1	0	0	1	0
4	0	1	0	0	0	1	1	0
5	0	1	0	1	0	1	1	1
6	0	1	1	0	0	1	0	1
7	0	1	1	1	0	1	0	0
8	1	0	0	0	1	1	0	0
9	1	0	0	1	1	1	0	1
10	1	0	1	0	1	1	1	1
11	1	0	1	1	1	1	1	0
12	1	1	0	0	1	0	1	0
13	1	1	0	1	1	0	1	1
14	1	1	1	0	1	0	0	1
15	1	1	1	1	1	0	0	0

A typical range of timing discs and timing strips.

interconnections that add greatly to the design of PCBs. With the development of inexpensive low power microprocessors this decoding can be implemented very easily and quickly by a small amount of software and converted by the same software if required into any other form of bit array such as 7 segment LED.

The commercial coding disc is relatively expensive to produce and may be etched metal, printed or photographically produced. An absolute encoder is one where at switch on the position can be read directly. There is another type of encoder that needs to travel through a home position at power up and then uses either offset sensors or offset holes to produce an up or down count. These are much less expensive to produce because using trigger on both negative and positive transition techniques, a 360 degree one degree resolution, timing disc can be made with only 181 slots. Triggering a clock on both negative and positive transition is a technique that uses two sensors. One is set to read a positive to negative transition and produce an output; the other produces a same polarity output on a negative to positive transition. The two outputs are 'or' gated effectively giving two outputs for each 'slot'. This type of disc is within the capability of most home PC printers to produce. This up down count technology is used on many ink jet printers and the head moves on power up to find the 'home' position. These printers operate 'on the fly', that is they use the optical strip for positioning and unlike a true servo system do a full traverse of the paper firing the ink as the print position is reached without stopping. Another application where this technology is used is the mouse on a computer. As the mouse is moved the PC moves the screen cursor to a matching position. The notion of a home position is used only in the context of the new home position i.e. where the cursor stops but, pixel and hence position count is relative to the screen.

Most early PC printers were of the dot matrix type that used a wire to impact a ribbon onto the paper. It was necessary for the print head to stop whilst the wire hit otherwise the result was torn ribbons, torn paper and bent wires. Stepper motors were generally used for positioning sometimes with a secondary optical step system for synchronisation. As the technology developed some later dot matrix printers used high-speed solenoids in the head drive that allowed printing on the move these normally used a DC motor for the drive and an optical strip for positioning. The optical strip was used for the 'dot' positions. Little advance was made on this front because these types of printers were being superseded by later technology such as ink jet and laser, which were capable of producing better graphics. The later dot matrix printers still persist in a number of limited applications such as ATM receipt printers.

Gray to binary conversion

This conversion follows a series of simple rules and steps.
1. Starting from the most significant bit (MSB) of the Gray code the binary bit is the same as the Gray bit up to and including the first '1' bit.
2. The Gray code bit now acts as a control.
3. If the Gray code bit is a '1', change the preceding binary bit to obtain the current binary bit. That is - if the previous binary bit was a '1' then the current binary bit is a '0', if the previous binary bit was a '0' then the current binary bit is a '1'.
4. If the Gray code bit is a '0'; the current binary bit is the same as the preceding binary bit. That is - if the previous binary bit was a '1' then the current binary bit is a '1', if the previous binary bit was a '0' then the current binary bit is a '0'.

Example

Consider the following example of a 10 bit Gray word converted to binary.
    Gray word 1011100010
The MSB is on the left preceded by a notional zero
    1011100010
This is the first '1' bit therefore the binary bit is 1
    1011100010
This bit is a '0' bit therefore the previous binary bit is repeated   1
    1011100010
This bit is a '1' bit therefore the previous binary bit is changed   0
    1011100010
This bit is a '1' bit therefore the previous binary bit is changed   1
    1011100010
This bit is a '1' bit therefore the previous binary bit is changed   0
    1011100010
This bit is a '0' bit therefore the previous binary bit is repeated   0
    1011100010
This bit is a '0' bit therefore the previous binary bit is repeated   0
    1011100010
This bit is a '0' bit therefore the previous binary bit is repeated   0
    1011100010
This bit is a '1' bit therefore the previous binary bit is changed   1
    1011100010
This bit is a '0' bit therefore the previous binary bit is repeated   1
Therefore by reading down the column
    Gray word   1011100010
    Binary word   1101000011

Binary to Gray conversion

This conversion follows a series of simple rules and steps.
1. Start from the most significant bit (MSB) of the binary number.
2. Compare each pair of succeeding bits.
3. If the bits are the same, put a '0' in the Gray code bit.
4. If the bits are different, put a '1' in the Gray code bit.

Example
Consider the following example of a 10 bit Gray word converted to binary.
    Binary word   1101000011
The MSB is on the left preceded by a notional zero
    01101000011
Binary bits are different therefore the Gray bit is 1
    1101000011

Binary bits are same therefore the Gray bit is 0
  1<u>1</u>01000011
Binary bits are different therefore the Gray bit is 1
  11<u>0</u>1000011
Binary bits are different therefore the Gray bit is 1
  110<u>1</u>000011
Binary bits are different therefore the Gray bit is 1
  1101<u>0</u>00011
Binary bits are same therefore the Gray bit is 0
  11010<u>0</u>0011
Binary bits are same therefore the Gray bit is 0
  110100<u>0</u>011
Binary bits are same therefore the Gray bit is 0
  1101000<u>0</u>11
Binary bits are different therefore the Gray bit is 1
  11010000<u>11</u>
Binary bits are same therefore the Gray bit is 0
Therefore by reading down the column
Binary word   1101000011
  Gray word   1011100010
These calculations are valid irrespective of the number of bits in the words.
Practical conversion circuits
Fig.104 shows a discrete component based Gray to binary converter using exclusive OR gates. This is for 4 bits but each extra bit only requires adding another XOR gate. The numbers indicate the bit weightings. Number 1 being the least significant bit.
Fig.105 shows a discrete component-based binary to Gray converter using exclusive OR gates. This is for 4 bits but each extra bit only requires adding another XOR gate. The numbers indicate the bit weightings. Number 1 being the least significant bit.

Servo position drives in the small workshop

The application for servo drive in the amateur or small workshop is limited. Most applications will be on/ off or full limit i.e. open or closed. Or in the case of a surface grinder or face milling operation may involve traverse to a limit switch position that turns off the drive motor or changes the direction.
Servo drives have always found application in models and for aerial rotation or set up. In the workshop they can be useful for adding controls for tools that are awkward to adjust from a work position or where the addition of extra wiring may be a hazard or infringe warranties. A mundane example could be a skylight or vent where different opening positions were needed. This could be operated by a simple DC motor using a two-way centre off switch for control and the operation watched until the desired opening was achieved. But if the vent could not be seen from the operating position then guesswork or a number of operate and go look at the results would be needed. Inversely the servo can be used as a means of feedback rather than of control. The simplest example of this is the fuel

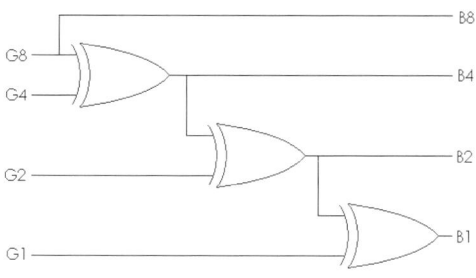

Fig.104

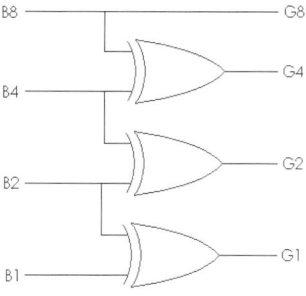

Fig.105

gauge in a car. In this application a float operates a variable resistance that in turn is used to balance the position of an indicator needle.

Simple servo position drive mechanisms

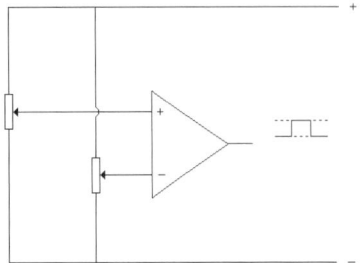

Fig.106 shows an OP amp used as the basis for a simple servo control. In this circuit when the voltage on the + input is greater than the voltage on the − input the output goes positive. The output goes negative when the voltage on the + input is less than the voltage on the − input.

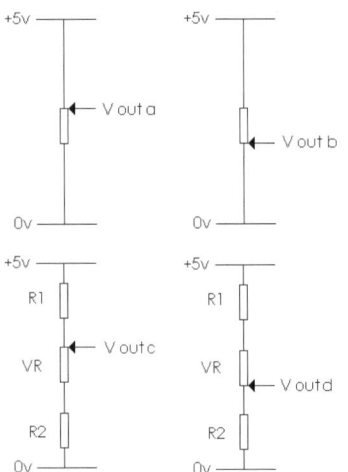

Fig.107 shows inputs to the simple servo system. Most OP amps can work rail to rail but you cannot get more positive than the + rail or more negative than the − rail hence switching near the extremes of travel when using the OP amp in voltage compare mode can be problematic. The diagram shows a possible solution to the problem. If V out a and V out b are from the feedback potentiometer the voltage swing is from 0V to 5V. If resistors R1 and R2 are placed in the set position potentiometer the voltage swing will be less than rail to rail. E.g. if R1 and R2 are each 2% of VR +R1 + R2 the voltage swing will be 96% of rail to rail i.e. from 0.1V to 4.9V. The exact set up and values chosen will depend on the mechanical and electronic components used. With the example given the voltage range of the feedback potentiometer exceeds the voltage range of the set position potentiometer and the system will therefore be able to reach an equilibrium point.

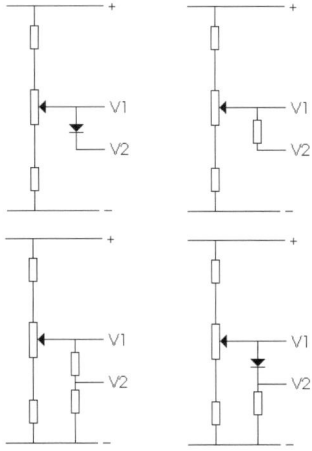

Fig.108 shows methods of achieving the 'dead spot' on the servo system. A diode has a constant voltage drop across it and can be used to produce the voltage differential required. The voltage will depend on the type of diode used but is typically 0.2 or 0.6V. The direct diode connection produces a linear shift whilst the potential divider method gives a non-linear movement but the voltage differential remains constant. Using a resistor in the potential divider method gives a differential that remains a percentage constant but the value varies with the applied voltage. The direct resistor connection is unlikely to work successfully because the low currents will mean that the voltage drop across the resistor is negligible.

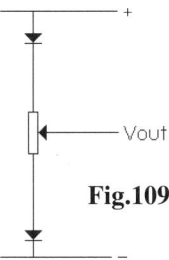

Fig.109

Fig.109 shows a method of using diodes to give a constant offset from the supply rails. A number of diodes can be used in series to produce offsets with steps of typically 0.2 or 0.6V depending on the type of diodes used.

One phenomenon that is little documented but can occur with OP amps used in voltage compare mode is equilibrium point oscillation. This is rapid switching of the output at the equilibrium point. Most of the cases I have come across are due either to bad circuit design or more often to poor PCB layout. All conductors have resistance although in practical terms it can be usually ignored when low currents are involved. One instance I was involved with of this having an effect was on a voltage comparator circuit that started a motor that ran through a sequence with the end result being the motor resetting the voltage comparator after it had moved to a second position. This is in essence a servo system. In tests a number of motors switched on and immediately switched off. After many long hours of testing and analysis it was found that because of the arrangement of connectors for the motor and for the voltage comparators the current taken by the motor at start up was sufficient to cause a voltage drop across the PCB +ve track that caused a voltage differential at the compare points opposite to and greater than the setting voltage hence the comparator 'reset'. The 'cure' was to use common point connections for the voltages supplying the points being compared and moving the motor supply point nearer to the voltage supply point on the PCB. These types of effects occur because of the sensitivity and speed of the OP amp inputs. One method of helping to prevent any low level oscillations caused at the switching point reaching the motor is to use a Schmitt trigger output on each OP amp or a set point switching made from spare OP amp circuits. If a small capacitor is used on the input of the Schmitt trigger it acts as a filter, smoothing out any input spikes and high speed switching. This will have a minor effect on the speed of response of the system.

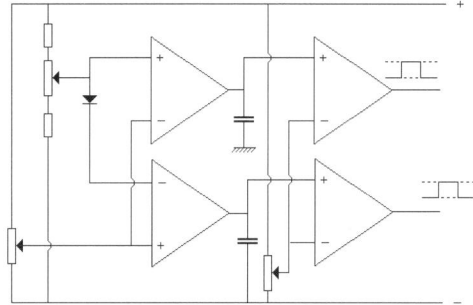

Fig.110 shows a circuit for a working servo system using the features discussed previously. The design is based on the LM324 quad OP amp run in comparator mode. The LM324 is designed to run from a standard +5V rail but will operate from +3 to +32V. The outputs are positive going and can be used to drive the two 'arms' of a 'H' bridge.

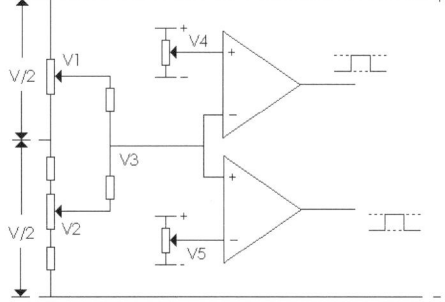

Fig.111 shows the progression from the previous circuit. It is probably the easiest to set up and most reliable circuit for this type of simple servo.

Above: Tacho section from a DC motor showing magnetic rotor and multi-pick up coil sensor.

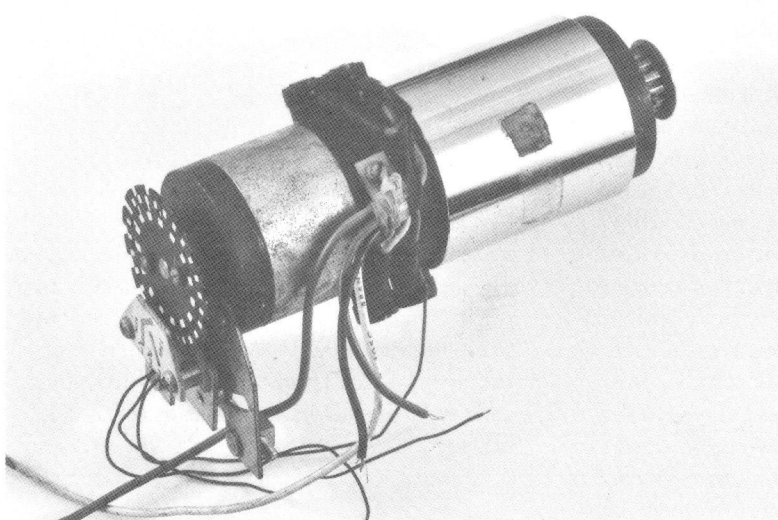

Left: A mid-range power DC motor with a tacho generator and optical direction sensing.
Above: A commercial motor with an integrated sensor block for speed control and direction sensing.

Right: A small sealed optical disc speed sensor shown complete and disassembled. The unit has a grating assembly producing moire interference lines that increase the count.

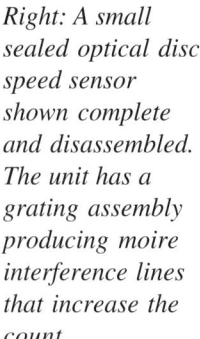
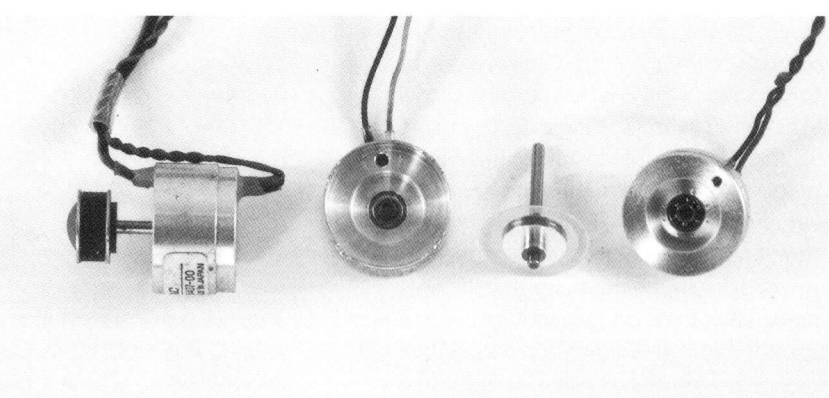

Electromechanical Building Blocks

OP1	OP2	OP3	OP4	Result
Off	Off	Off	Off	No output or motor 'coasting' (no braking)
On	Off	Off	On	CW rotation
Off	On	On	Off	CCW rotation
Off	On	Off	On	Braking

This table shows the outputs needed from a circuit when an 'H' bridge is selected.

OP1	OP2	OP3	OP4	Result
Off	Off	Off	Off	No output or motor 'coasting' (no braking)
On	Off	Off	Off	CW rotation
Off	On	Off	Off	CCW rotation

This table shows the outputs needed from a circuit when the relay output is selected.

All the principles discussed previously also apply to this diagram. The set position potentiometer producing V2 and the motor positioning potentiometer producing V1 are placed in series across the power rails. The junction of these will be approximately V/2 but will depend on the offset resistors on the set potentiometer to ensure the range of V1 > V2. The two extra resistors that give V3 produce a type of bridge circuit with the potentiometers. When the potentiometers are at centre position V3 ≈ V/2. If the voltages V4 < V3 and the voltages V5 > V3, then both the outputs will be switched off. When V4 > V3 the output associated with V4 will be on. When V5 < V3 the output associated with V5 will be on.

Set up is a number of simple steps. Measure the voltage on the ends of each potentiometer and set V1 and V2 to the mean of their corresponding upper and lower voltages. Measure V3 this should be approximately V/2. Set voltage V4 to be V3 − Vdz/2 and voltage V4 to be V3 + Vdz/2. Vdz is the required dead zone voltage width. This means that at the null point a change of Vdz/2 will be needed before either of the motors will be switched on.

Practical values are not critical and for this type of circuit I find the following values work correctly.

Set position potentiometer = motor positioning potentiometer = 47k
Offset resistors = 1k
V3 resistors = 100k
V4 potentiometer = V5 potentiometer = 100k

Depending on the positioning of the set position potentiometer and motor positioning potentiometer noise can be a problem. It is simple to add an OP amp configured in differential mode to reject common mode noise. For greater distances the set position potentiometer can be replaced with a digital potentiometer.

The circuits described are all 'bang – bang' types. This means the motor has three states, full speed clockwise - stop - full speed anticlockwise. More efficient circuits use proportional control to change the motor speed in relation to the present position and the null point position. These can be made using variable gain amplifiers to a pulse width motor driver. Feedback from the motor current can be also used to give a degree of proportional speed control. But since the advent of the small microprocessor it is not really worth the effort to produce this sort of analogue circuit when the control can be implemented so simply using digital techniques. It is also possible to add features such as hard or soft braking using a digital driver.

The previous circuits can be used with 'H' bridge drive or two relay direction drive as described in the motor drive circuits. If relay direction circuits are

The Servo System

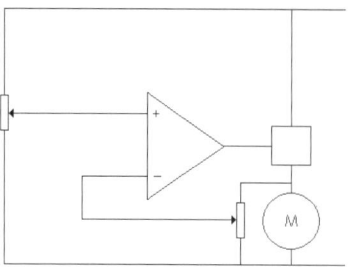

Fig.112

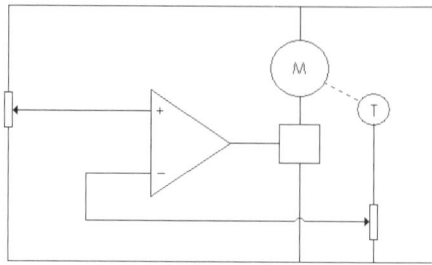

Fig.113

used with this type of position servo circuit the time for the relay to 'break' must be taken into account and will probably affect the width of the 'dead zone' required. Braking will be automatic when the activated relay is released.

If relays are used the braking outputs can be ignored because braking will be automatic as the relay opens using the two relay method. The use of relays or 'H' bridge drivers is a selectable option.

Servo speed motor drives

There are a number of circuits used to control the speed of DC motors that use the comment 'this diagram offers some degree of speed control'. This can usually be translated to mean that the speed control is in a wide band. Accurate speed setting at a fixed point or over a very narrow band is relatively easy using digital measuring techniques. The wider the speed band required the more difficult the problem becomes. A small DC motor may have a full speed in excess of 20000 RPM. Permanent magnet and parallel wound field motors are easier to control because of there inherent torque/ speed characteristics. Many simple attempts at speed stabilisation are based on measuring the current through the motor by measuring the voltage across it or across a low value resistor. Speed will drop with increasing load and the current will increase. Increasing the voltage will increase the speed and therefore reduce the current consistent with the load.

The most common commercial methods are firstly to count a number of pulses within a given time frame and compare this with a set count the motor output drive is adjusted to bring the counts to match. The second method is to use a tacho-generator that produces a voltage that is proportional to speed. This voltage is compared with a set voltage and the motor output drive is adjusted to bring the voltages to match. This second method is an analogue method and can be treated in a similar way to a position servo.

Fig.112 shows a feedback method where the voltage across the motor switches off the output when the voltage reaches a set level and switches back on when the voltage falls below the set level. The output is in effect a 'chopped' voltage.

Practical servo speed motor drives

Fig.113 shows a feedback method for a motor turning in one direction only. The voltage across the motor is switched off when the voltage from the tacho-generator reaches a set level and switches back on when the voltage falls below the set level. The output is in effect a 'chopped' voltage. This method is independent of current flow in the motor and relies totally on the voltage from the tacho-generator.

The individual component values will depend on the motor voltage and the tacho output. The LM324 will work for the voltage comparator and the driver can be chosen from the low side MOSFET circuits described earlier.

In the same way that position servos use proportional control in the better designs then better designed speed servos use proportional control in the form of PWM for speed control. One disadvantage of chopper type control is the

Motor with slotted optical disc and offset sensor pair capable of speed and direction sensing.

possibility of interference that can be put onto the logic and power lines and very careful design is necessary especially when using sensitive high speed controllers in a system because the chopping can be picked up as changes in logic levels if steps are not taken to suppress interference before it is induced into cable runs.

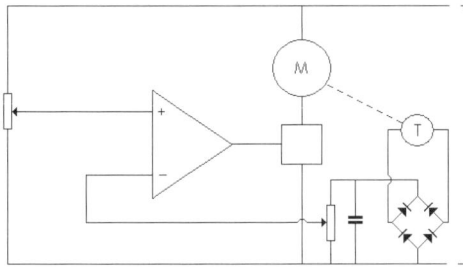

Fig.114 shows a feedback circuit that will run in either direction by using either a change over relay or a 'H' bridge circuit for the motor. The bridge rectifier produces an output that is positive with respect to ground irrespective of the polarity of the output from the tacho.

Fig.115 shows a theoretical feedback circuit that will run in either direction by using either a change over relay or a 'H' bridge circuit for the motor. The control in this case is by using the feedback from the tacho to control a PWM

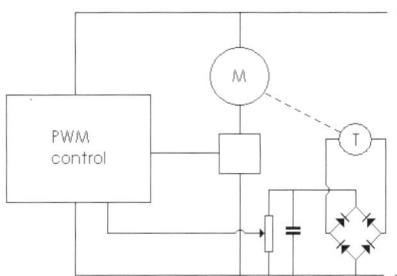

Fig.115

circuit. A simple form of PWM controller with a limited range can be made from the 555 type circuits described earlier with the feedback being used to set the control voltage term on pin 5. Depending on the speed range it may be wise to fit a zener diode between the potentiometer slider and ground to protect the controller in case of errors occurring that place a higher than normal voltage on the controller pin.

Fixed range drives

There are many applications where a motor is needed to traverse a specific part of a rotation and stop at a fixed point with the option to then return to the start point e.g. a motorised valve. This could be achieved using a stepper motor but the application rarely warrants the expense. A small geared motor is the obvious answer.
It will usually be necessary to brake the motor at each required stop position. With a small amount of control logic it is also possible to provide a number of intermediate positions.
The normal control is by using one or more cam plates with the halt positions set in them. The sensing method can be micro switches, magnetic sensors or optical sensors.

A discrete component fixed range motor drive with auto return

This unit was designed to open a valve when a signal causes the relay to close, and removal of the on signal causes the valve to close automatically, it does not require a close signal. The control system uses limit switches for both

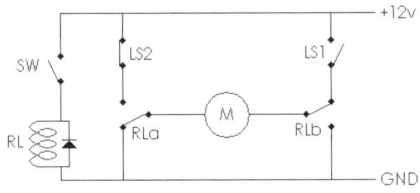

Fig.116

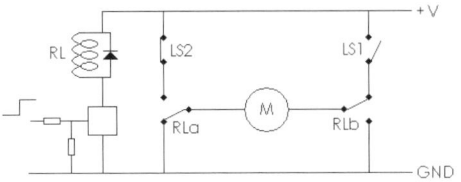

Fig.117

the fully open and fully closed position.
Fig.116 shows a simple form of this circuit using a switch to control the relay.
When the relay is operated RLa and RLb will transfer. +12volts will flow via LS2, RLa, the motor and RLb to ground. The motor will turn and the valve will open. When the valve reaches the required CW position LS2 will open and the motor will stop. When the relay is released LS1 will be closed. Therefore +12volts will flow via LS1, RLb, the motor and RLa to ground. The motor will turn CCW and return to the rest position. When the motor reaches the rest position LS1 will open and the motor will stop.
This is shown with relays and switches but could as easily be achieved with other motor drivers and sensors in place of the limit switches. The voltage was chosen to match the motor and relays but other voltages are equally feasible.
Fig.117 shows a block diagram for a MOSFET driven relay working as the previous switch controlled unit.
Fig.118 shows the block diagram of a discrete

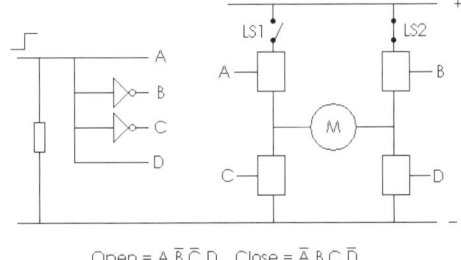

Open = A $\bar{B}$ $\bar{C}$ D   Close = $\bar{A}$ B C $\bar{D}$

Fig.118

component based circuit to produce a simple fixed range motor drive using 'H' bridge motor drivers.
When a positive input is seen, A= 1, B = 0, C = 0 and D = 1. Drivers A and D will turn on and the motor will rotate until limit switch LS1 opens cutting power to the motor.
When the positive input falls, A= 0, B = 1, C = 1 and D = 0. Drivers B and C will turn on via the inverters and the motor will rotate until limit switch LS2 opens cutting power to the motor.

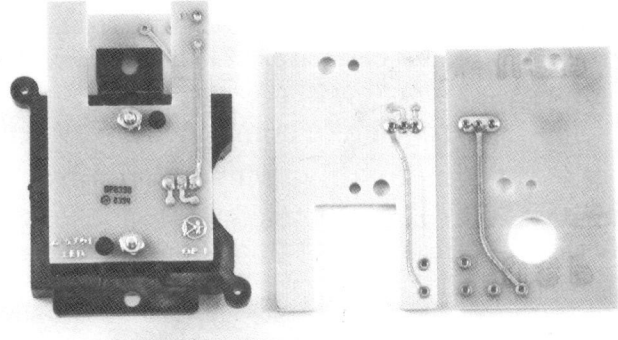

Early type of speed/ position sensor using three sensor and LED pairs

CHAPTER SIX

Types of Relay

Relays are devices that allow switching of electrical circuits by the use of a magnetic coil. Large relays usually require either 12 or 24V DC or 240V AC. Relays may have multiple combinations of switching options.

Normally closed/ normally open

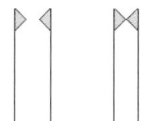

Fig.119 shows the simplest type of relay containing one set of contacts that either closes when the relay is activated or open when the relay is activated. These will be designated respectively normally open (NO) and normally closed (NC).

Changeover

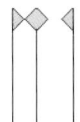

Fig.120 shows the contacts of a changeover relay. The contact can be in one of two possible states. When the relay is activated the centre contact will move from normally closed (NC) contact position to a normally open (NO) contact position. This type of contact set is designated changeover (CO).

Centre off

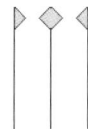

Fig.121 shows the contacts of an unusual relay that is expensive to produce and is now normally replaced by two simple standard relays. The relay consists of a centre armature carrying a set of contacts. The rest position is the centre. The armature can be pulled left or right by activating one of the two coils. These are fixed at each side of the armature. Activating one of the coils pulled the centre contact to the matching outer contact.

Solid state

The conventional solid state relay based on a triac or bi-directionally mounted SCRs do not work with DC inputs because once switched on, they only switch off when the voltage across the

A solid state relay for mains output and low voltage DC control.

load falls to zero i.e. when the AC sine wave crosses the zero level.

There are a few devices about are based on high voltage components such as IGBTs that can handle high voltage DC. These are not relevant to this book because it is unlikely that the voltage used with the type of devices being driven will exceed 50V.

The one part of the SSR that can be carried over is the optically isolated input that prevents any risk of high voltage reaching low voltage logic lines. It is particularly useful if the driving unit is a PC. This is discussed on the section on optical isolation.

Reed relays

These are glass tubes containing a pair of blades that under the influence of a magnetic field either from a permanent magnet or from a coil close to complete a circuit. The most common format is one normally open (NO) contact. There is a less common changeover (CO) version. The power and voltage rating of full size reed relays rarely exceeds 2A and the miniature versions substantially less. Winding a coil to go around a number of reed relays can produce multiple contact options. There are also versions intended for use in electronic circuits that are fully encapsulated in a SIL or DIL package with a coil built in.

Coils are easily wound for use with reed relays. The information needed to enable this to take place is the amp/turns. This information is normally available on the manufactures data sheet but failing this can be worked out by trial and error. A good starting point is 80 for standard size and 25 for miniature. This means that in the case of the standard size, 1 turn would need 80A to operate and 80 turns would need 1 amp to operate.

In practical terms I would probably overrate this by about 50% and in this example use 120A turns in round figures for calculation purposes.

Worked example

Available voltage 12V
Amp turns 100
Mean diameter of coil 2.5cm (chosen to fit relay)
Working temperature 20°C
The above items are fixed by physical factors.
A choice was needed about the desired current and in this case 100mA was chosen because this is the normal current of many relay coils and is well within the specification for a small driver. From ohms law it is easy to calculate the resistance needed for the circuit. This is a practical circuit and no allowance was made for the resistance of the driver. This is very small compared to the coil resistance.
$R = V/I$ i.e. $R = 12/0.1$

R = 120 W
With a current of 0.1amp and a coil needing 100 amp/turns we therefore need 1000 turns.
The coil has a mean diameter of 2.5cms therefore the mean circumference is 2.5p this approximates to 7.85cms
1000 turns of 7.85cms approximates to 78.5 metres
The required resistance is 120 W
This equates to a resistance of 1.52W/metre
From a wire resistance table (see Chapter 23 Pin Outs & Specifications)
36 gauge wire is 1.32W/metre and 38 gauge wire is 2.11W/metre
Neither of these are an exact match therefore I would work on the basis of the 36 gauge wire
This recalculates to 91 metres for the required resistance
91 metres is 1160 turns
The amp/turns (1160 x 0.1) equates to 116 this is within the allowance of error.

Relay ratings

Relays are rated on their current switching capability. This is the resistive load and the inductive load is a fraction of the resistive load. Typically, it may be only one tenth. Relay contacts can produce arcing and a large relay may typically have a close time of 20mSec and an open time of 5mSec. Relays rated at 240V AC 10A are often only rated at 30V DC for the same current. The same relay may only be rated at 240V AC for ½ HP inductive switching. Therefore by implication is approximately 1.5A at 240V AC. The same will apply at 30V DC inductive load. Therefore the switching capability is only 90 watts i.e. approximately 1/8 HP.
There are available a large range of high current 12V relays available that are designed for automobile use.
With correctly designed circuits, the relay will not be used for switching outside of the rating. Therefore effects such as contact weld due to arcing will not occur.
There are a number of factors that affect the suitability of a relay for a particular project. These include contact material, contact separation, insulation, open/close time temperature rating and current rating. This information is readily available from the manufacturer or any reputable supplier.

Mercury wetted relays

These tend to be used for fixed position high voltage and high current switching. Ratings are typically 230Vac and 45A. The mercury coats the switch contact faces and because it is liquid the contact is very low resistance and self-renewing. Some smaller versions were used in telecommunications before electronic became the norm.

Time delay relays

This is essentially a standard relay to which is added an electronic control circuit that holds the relay in for a period of time when initiated. There is also a version that switches off for a set period. They are used mainly for process control. Because of the limited application they tend to be about five to ten times the cost of an equivalent relay. Some of the timers are built in to base units with relay sockets. The timer mechanism is easy to produce electronically as an add on unit.

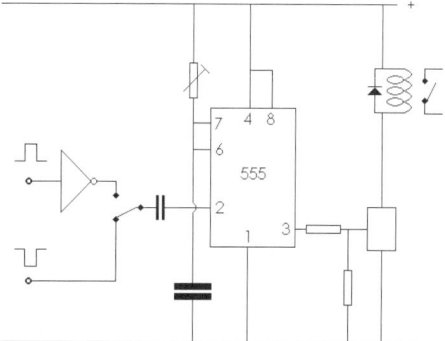

Fig.122 shows a simple circuit based around a standard driver and a 555 monostable with an inverter if positive triggering is required. The circuit is capable of holding the relay in from seconds to many minutes but very long timescales may be inaccurate due to leakage in the timing capacitor.

For long delays a dedicated long delay IC such as the ZN1304 or a counter and divide circuit can be used.

For a switch off for a time period circuit use a normally closed relay contact set.

Transistorised relays

These are relays able to be switched by low power inputs but have high power output. They are essentially a standard relay with a Darlington or MOSFET driver built in. They may also have a voltage regulator circuit that means they can be used on an extended voltage range.

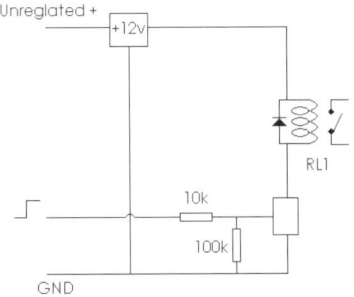

Fig.123 shows the simplest form of single relay driver. It is possible to build this into a small box fitted with a relay socket. The regulator allows connection to unregulated voltages of up to approximately 30V. This regulator could be substituted with a +24V regulator for 24V solenoids. An LED could be fitted in parallel with the relay if required. The LED resistor would need to match the regulated voltage.

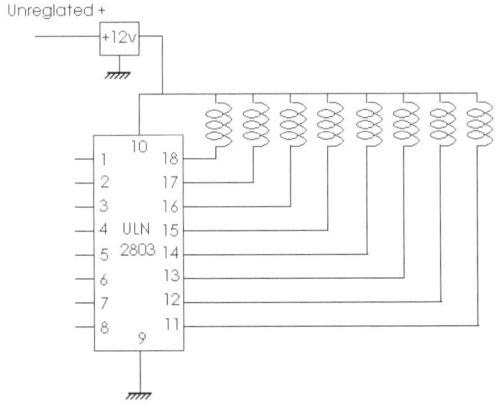

Fig.124 shows a set up that allows up to eight relays to be switched from positive going logic. The circuit requires no freewheeling diodes, input limiting resistors or input to ground resistors because these are all built into the ULN2803.

The power voltage input can be up to 50V therefore a range of regulators can be used to match the relays that must all be the same voltage rating.

Spike suppression

The relay like any wound component produces a collapsing field when voltage is removed. A back EMF suppression diode will be required to prevent damage to the driver unless it is contained within the driver e.g. the ULN2003. Most relays use relatively small currents therefore any small fast diode with a current rating to match the relay coil will suffice.

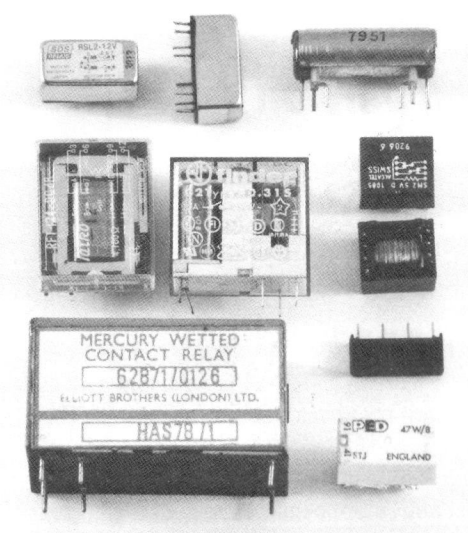

A range of low current signal and control relays.

CHAPTER SEVEN

Solenoids & Related Devices

Types of solenoids

A solenoid basically consists of a coil with a hole passing through it longitudinally inside an iron frame. When current is passed through the coil an electromagnet is created and a loose plunger is pulled into the coil. The plunger being shaped with a conical end and a matching iron section forming part of the frame but also being inside of the coil enhances the pull.

It may be necessary to fit anti-residual springs to move the plunger back to the open position if the unit the plunger is connected to does not provide the necessary pull.

Pull solenoid

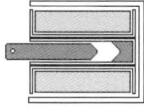

Fig.125 shows the commonest type of solenoid. It is the solenoid type that the standard description is based on. There is a large number of types and specifications.

There is a less common type usually produced to provide high power. This consists of a square plunger with two lugs. The frame is heavier than the standard type of solenoid and has frame sections that match with the lugs. The effect is reminiscent of a horseshoe magnet, the attraction occurring by the square section being drawn into the frame but also between the frame and the lugs.

Push solenoids

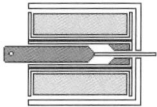

Fig.126 shows a push solenoid. These are often of the standard pull solenoid types with an extension of the plunger protruding through the end of the frame. The size of the plunger is limited in relation to the plunger diameter. Therefore high power push solenoids consist of a pull solenoid with a leverage system built onto the frame.

Centre off

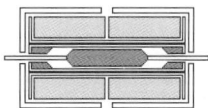

Fig.127 shows a centre off solenoid. This is in effect a plunger with a push extension on each end inside a frame with a centre tapped or two separate coils. The solenoid can be used in tri-state or bi-state mode i.e. if the plunger is fitted with centring springs it will return to the centre from whichever end it was at when power is removed. Without the springs and with no other

A range of solenoids.

force to move it to the centre it will remain at whichever end it was at when the power was switched off. This type of solenoid may be used for interlocking to prevent access to two parts of a unit at the same time. Small permanent magnets may be used on the bistable type to prevent the plunger moving due to vibration.

Variable positioning solenoids - balanced

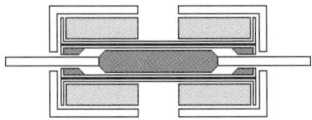

Fig.128 shows a variable position balanced solenoid. The difference between this and the centre off solenoid is that the coils in the variable positioning solenoid are both on at the same time. These use a similar type of technology to microstepping for stepper motors. The unit consists of a core with an extension for attachment to the driven mechanism sitting between two solenoid coils. The current through the coils can be balanced to adjust the position of the core relative to the coils. They are often used in process control to operate the opening of valves. The units tend to be large and may have forces in the order of ten kilograms.

Variable positioning solenoids - ratchet

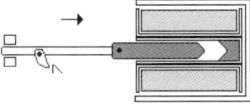

Fig.129 shows a variable position ratchet solenoid. These use a pulsed coil and a ratchet and pawl or similar drive such as a one way roller or spring clutch to make a stepped physical movement. They are one way devices but resist being turned backwards when power is removed. The most prolific example of these was probably in the Strowger switchgear used in early pulse drive telephone systems. Each pulse caused the solenoid

A commercially produced unit showing rotary movement from a straight pull solenoid.

to move one position moving a set of contacts to complete the telephone circuit to the required line.

Rotary solenoids

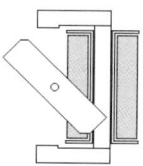

Fig.130 shows the first type of rotary solenoid. It is a true rotary solenoid and is similar to a DC motor without the commutator. The coil may be on the core or on the moving section and connected by flexible leads. A small motor can be modified to produce a rotary solenoid. Another method is to use a magnet set out of alignment with a field winding. When power is applied the magnetic field attracts the magnet. This type of solenoid usually requires a return spring because of the attraction between the magnet and the pole pieces when no power is applied. The collapsing field that is opposite to the on field also creates a repulsion of the magnets but only for a very short time. Very small rotary solenoids may use an iron rotor similar to the variable reluctance stepper motor instead of a magnet. Multiple coils can be used to produce both attraction and repulsion to achieve greater

torque in permanent magnet types.

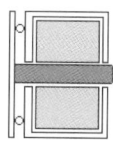

Fig.131 shows the second type of rotary solenoid. This uses a straight pull on a plate that has a number of incline slopes and ball bearings. The effect is that the pull causes the ball bearing to run along the incline slope and the effect is rotary motion.

Rotary movement from a straight pull solenoid

When greater power is required a straight pull solenoid can be used with a toothed belt or cable running over a pulley or shaft and attached to a

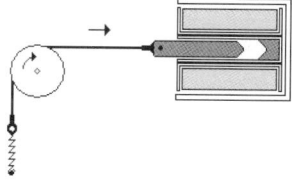

spring. **Fig.132** shows a typical example. When the solenoid is activated the belt or cable is pulled causing the shaft to turn. This technique is also used in disk drives for moving the head

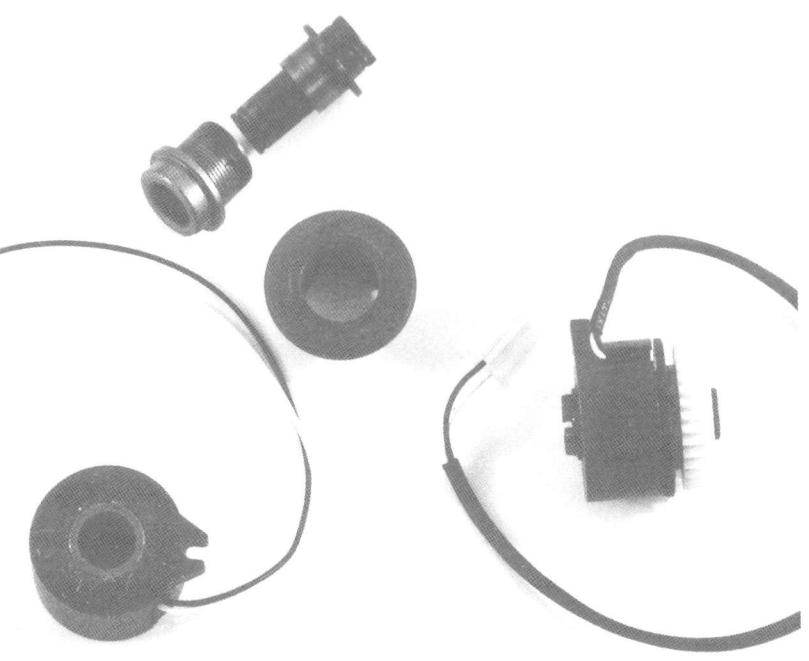

A small electro magnetic actuated spring clutch shown complete and disassembled.

over the disk but using a stepper motor or 'voice coil' positioning solenoid.

Magnetic latching solenoids

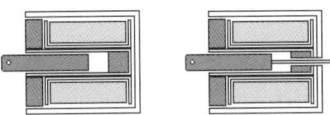

Fig.133 shows a magnetic latching solenoid. This is a solenoid with a built in permanent magnet that causes the plunger to stay in the close position after activation and removal of the power. An opposite polarity pulse will negate the effect of the permanent magnet. The pulse does not provide power to move the plunger out of the coil to its open position, the system it is attached to or a spring must do this. The coil may be a split coil, two separate coils or a single coil. Depending whether the solenoid has two separate coils or a single coil will determine if it is driven by single ended drivers or half of an 'H' bridge. This is a similar concept to driving unipolar and bipolar stepper motor coils.

High power solenoids

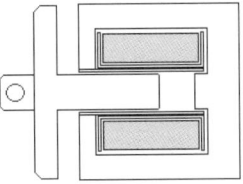

Fig.134 shows a high power solenoid. The basic design is the same as a standard pull solenoid except for the shape. The core is shaped in the form of a horseshoe magnet. The plunger has extensions that match the end faces. It is available in DC and AC versions. The AC versions generally have laminated bodies and cores to minimise eddy currents.

Difference between AC and DC solenoids

Although this book is mainly dealing with DC

solenoids it is desirable to understand the basic differences between AC and DC solenoids because there is a limited range of low voltage AC solenoids available. With a DC solenoid the current rise at switch on will rise until the current 'matches' the DC resistance. The inductance of an open solenoid is less than the inductance of a closed solenoid but the current cannot exceed that set by the DC resistance. In a solenoid the current rise that occurs at switch on is termed the inrush current. See section on current rise in inductors.

In an AC solenoid the inrush current occurs at twice the frequency of AC supply i.e. once on the positive cycle and once on the negative cycle. In an AC system the current limit is a function of the reactance and hence the inductance. Because the inductance is less in an open solenoid than in a closed solenoid the current will be much higher in the open solenoid. This produces three effects. The first is that an AC solenoid tends to be more powerful in the open position than the equivalent DC solenoid. The inrush current on an AC solenoid may be ten times that of the closed position current. The second effect is that an AC solenoid is generally faster in operation than the equivalent DC solenoid depending on the start position. The third effect is that an AC solenoid that does not close fully is likely to overheat and possibly burn out. Methods can be used that bring the characteristics of a DC solenoid up to that of the AC solenoid without the risk of burn out. See section on high voltage driving and increasing drive power from low power circuits.

Solenoid valves

These are as the name suggests a solenoid type coil operating a liquid or gas valve. The plunger is usually inside a tube that forms part of the gas or liquid flow route. A variation on this uses a plunger seal inside a set of flexible bellows.

Solenoid ratings

All solenoids have manufacturer ratings but the information may vary with the information given and how it is presented. The following is a typical set of characteristics that may be found.

Voltage
This is the rated voltage at the stated duty cycle.

Wattage
This is based on maximum allowed power based often on the duty cycle and the working temperature range.

Temperature
This may be given in relation to ambient temperature and the maximum working temperature.

Force/ stroke curve
This is usually presented in the form of graphs based on the working voltage. The graphs show the force over the designed movement range of the plunger.

Operate time
At any point in the force/ stroke curve the difference between the force required to move the load and the force supplied by the plunger is the available force to accelerate the plunger. Increase in load will affect the operate time. Overrating a solenoid produces a larger excess force but the energy of the plunger will need to dissipate when the solenoid fully closes.

Holding force
This is the force necessary to 'break' the hold of a fully closed solenoid.

Duty cycle
This is probably the most complicated of the solenoid characteristics. The duty cycle is usually presented in the following form.

Rating
This is normally shown as continuous (100%), 50% or 25%. This is the time a solenoid can remain on at stated conditions. This means that a 100% rated solenoid can remain on continuously without exceeding the working temperature. This is often based on an ambient temperature of 20°C because the higher the ambient temperature the less will be the thermal gradient and the slower the heat will be lost from the body of the solenoid. Some solenoids have ratings of higher than 100%. One particular solenoid I used frequently had a rating of 140%. This meant it

A commercial 180 degree half turn spring clutch. The separate springs are drive and anti-reverse.

could be easily run at higher voltages or at its rated voltage there was little risk of failure due to overheating etc.

Consider a solenoid capable 12V operation at 5 watts. It could be operated on a 25% duty cycle at 20 watts.

With calculations taken to the nearest single place of decimals.

This means that at 100% cycle the current from $P = IV$ would be 0.4A.

Therefore from $V = IR$ the DC resistance of the solenoid is 30W.

Therefore to produce a 20 watt rating the current is 0.8A and the voltage needed is 24V.

This could have been calculated from the fact of doubling the power doubles the current and hence increases the power by 2^2.

This is $5 \times 4 = 20$ watts.

Maximum on time

Duty cycle by itself is meaningless except for continuous rated solenoids.

If a solenoid was run at 1 minute on and 3 minutes off this is a duty cycle of 25%. If the same solenoid was run at 1 hour on and 3 hours off this is also a duty cycle of 25%. Therefore a maximum on time will also be stated. A typical maximum on time may be in the order of 2 minutes. This means that the 25% duty cycle would be 2 minutes on and 8 minutes off. If the maximum on time in the design of the circuit was less than the maximum then the calculations can be based on pro rata times.

Complications occur when this information is not given or the design requires a solenoid to be used in a manner that does not conform to a regular cycle. In this case it may be wise to use a solenoid based on the 100% rating or monitor carefully the temperature when testing the design.

High voltage drive

The solenoid can be treated in the same way as the stepper motor coil with a high voltage being placed on the solenoid initially and the current flowing measured or an elapsed time period or the solenoid position can be used to drop the voltage to a hold level suitable for the solenoid. This technique often allows a smaller solenoid to be used because the power is increased over the area of weakest pull of the solenoid. This is a way of obtaining substantially more power at the beginning of the force/ stroke curve without

A 40 year time span of air, vacuum and liquid solenoid valves.

causing a great effect on the duty cycle. If the solenoid is used at a fast cycle rate it will be necessary to consider the percentage of the time that the voltage is at a high level compared to that a normal level and make allowances in the design.

Increasing drive power from low power circuits

It can be a problem driving solenoids from battery circuit when the internal impedance of the batteries cannot provide the necessary current that the solenoid requires.

The best solution is probably to replace the batteries with NiCd/ NiMH or sealed lead acid batteries.

If this is not feasible and the on time of the solenoid is short a capacitor can be used. The capacitor has a very low internal resistance and can provide current almost instantaneously. The capacitor is charged slowly from the batteries and when charged is discharged through the solenoid. The technique is particularly successful with magnetic latching solenoids. If the battery potential is low, voltage-multiplying techniques can be used to put a higher potential on the capacitor. All electrolytic capacitors suffer from leakage current and an electrolytic capacitor will discharge over time and also flatten the battery eventually if it is left in the battery circuit permanently.

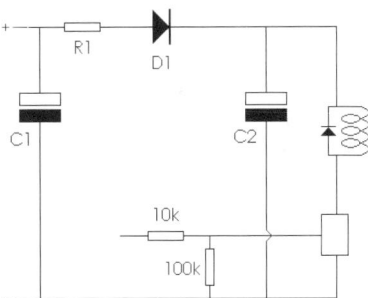

Fig.135 shows a circuit for use with a solenoid from a battery circuit. The capacitor C1 is to help prevent the main battery circuit being pulled down when the solenoid is operated. The resistor R1 limits the current flow from the battery to the capacitor C2 and this again helps to prevent the main battery circuit being pulled down when the solenoid is operated. The diode D1 prevents the capacitor C2 discharging

Section of a milling machine leadscrew drive unit showing timing disc and sensor and solenoid driven dog clutch assembly.

through the rest of the circuit. The charged capacitor C2 provides the power to operate the solenoid. The driver circuit is standard and the driver must be capable of carrying the current of the discharging capacitor. To calculate this, assume C2 is low impedance source and has no resistance to the supply of current. The resistance can be assumed to be the DC resistance of the solenoid coil. This will give a guide to current rating necessary.

Fig.136 shows a similar circuit but with a voltage multiplier or booster to charge the main capacitor C2.

Fig 137 shows a circuit that is useful for a number of solenoids. The advantage is all the power drivers are low side making the driving simple and not requiring any boosted drive. The circuit uses two power drivers, one connected to the 'standard' ground connection and the other to an extra supply that is negative in relation to the 'standard' ground. If the solenoid is 12V rated and the main supply is +12V level and the negative supply is –12V level, then the option is a supply to the solenoid of either +12V or +24V. If MOSFETs are used for the drivers, care must be taken with the drive voltage levels. If the +12V line is used as a source of drive for input 1 there is no problem because the drain to source voltage will be +12V. If the same level is used to drive input 2 this is equivalent to a drain to

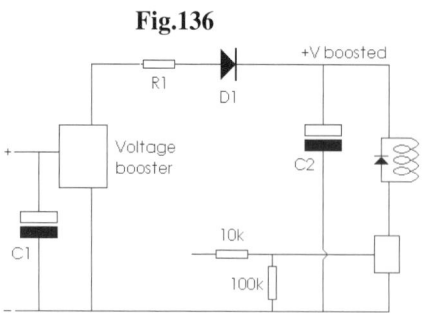

Fig.136

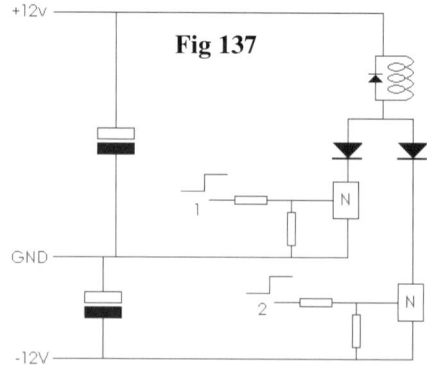

Fig 137

Electromechanical Building Blocks

Disassembled solenoid valve.

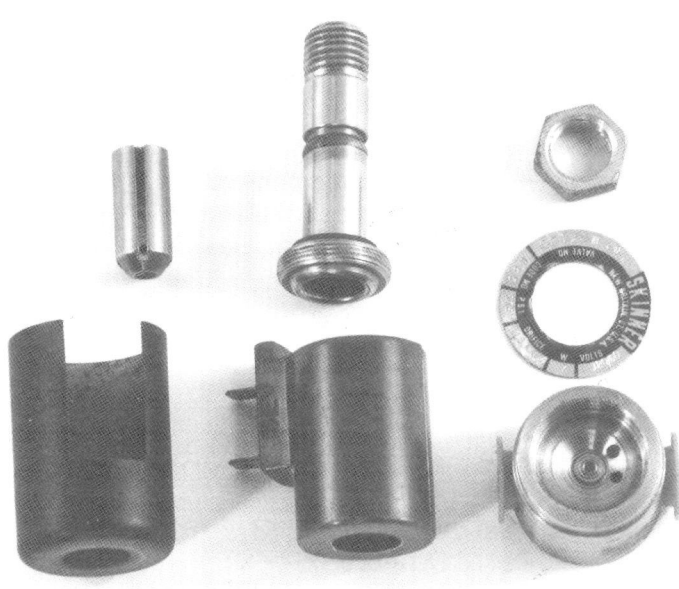

source voltage of +24V. Zener diodes or voltage dividing techniques should be used.

The +24V supply would only be used for a short period when the solenoid is turned on, and would therefore not need a supply capable of supplying high current continuously. A charged capacitor could be used as in previous circuits dependant on the cycling times of the solenoids to provide a low impedance current source.

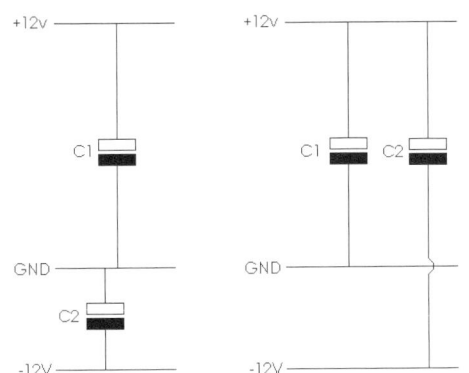

Fig.138 shows two methods of wiring in the capacitors in the circuit. Both methods use C1 to supply current when 12v drive is selected. With the first method C2 is effectively in series with C1 and would therefore only be of the value for series capacitors i.e. $1/C = 1/C1 + 1/C2$. Each capacitor would only need to be rated for the voltage across it, in this case 12V plus chosen margin.

In the second case independent capacitors are used for the two switching levels. This means that a low impedance current source is also available at the transition of the voltage level switching. C2 in the second version will need to be rated for the voltage across it, in this case 24V plus chosen margin i.e. $((+12) - (-12))$. The size of capacitor needs to be chosen such that it can be fully charged by the power supply being used between solenoid cycles.

This technique can also be used with small motors to produce a starting surge with motors that are loaded and require a start up boost to overcome inertia and drag.

A discrete component circuit for driving a solenoid boost

This area describes the use of a discrete component based circuit to produce a simple and practical circuit for driving a solenoid boost.

Fig.139 shows a practical circuit for providing a short boost drive when a solenoid is turned on. The input pulse turns on the drive to the solenoid and also triggers a monostable based around a 4047. The capacitor resistor network around pin 9 ensures that the device is reset at power up and prevents the solenoid firing at switch on. It may be necessary to modify the timing of this depending on the type of power supply being used.

C1 and R1 provide the monostable timing. C1 must not be a polarised type. The output timing is 2.48CR seconds. As examples if C1 equals 1mF and R1 equals 470K the output is 1.17 seconds. If C1 equals 0.47mF and R1 equals 470K the output is 0.55 seconds.

Out of phase outputs are available depending on the drive polarity required i.e. negative going positive or positive going negative.

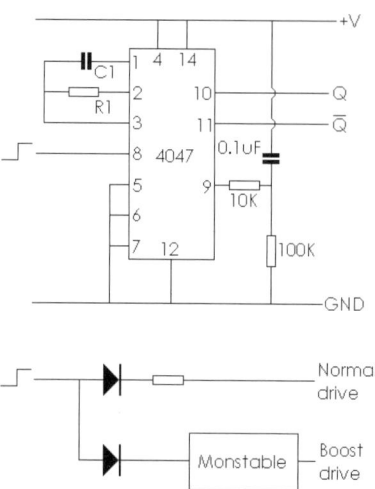

Fig.139

CHAPTER EIGHT

Other Electromagnetic Devices

Wrapped spring continuous rotation

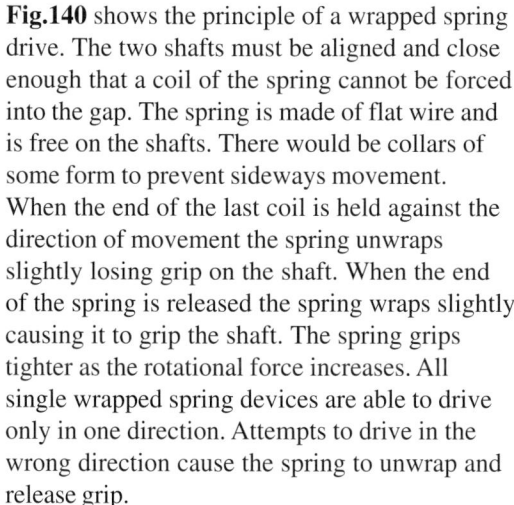

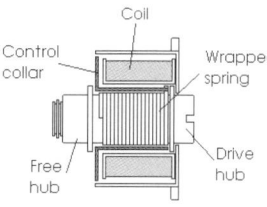

Fig.141

Fig.140 shows the principle of a wrapped spring drive. The two shafts must be aligned and close enough that a coil of the spring cannot be forced into the gap. The spring is made of flat wire and is free on the shafts. There would be collars of some form to prevent sideways movement. When the end of the last coil is held against the direction of movement the spring unwraps slightly losing grip on the shaft. When the end of the spring is released the spring wraps slightly causing it to grip the shaft. The spring grips tighter as the rotational force increases. All single wrapped spring devices are able to drive only in one direction. Attempts to drive in the wrong direction cause the spring to unwrap and release grip.

Fig.141 shows the typical layout of continuous rotation wrapped spring drive. When the magnetic field is activated the control collar is attracted to the coil and in the process causes the control hub to resist turning and the wrapped spring grips the driven hub and the free hub, effectively joining them together and causing rotation. The drive is usually transmitted onwards by a gear or belt drive on the free hub. The drive can be halted at any position on the rotational cycle by releasing the magnetic field.

Wrapped spring single rotation

This is not a true electromagnetic device. In reality it is a mechanical device operated by a solenoid. When the solenoid is activated a detent or release arm is pulled away from a stop on the release collar. The collar turns under pressure of the wrapped spring and the wrapped spring grips the shaft causing it to be connected to the drive. The unit rotates until the detent is dropped into the path of the collar stop. The unit continues to rotate until the detent pushing against the collar stop causes the wrapped spring to disengage. This type of drive is capable of driving heavy loads. It will always stop at the same rotational position hence the name single rotation.

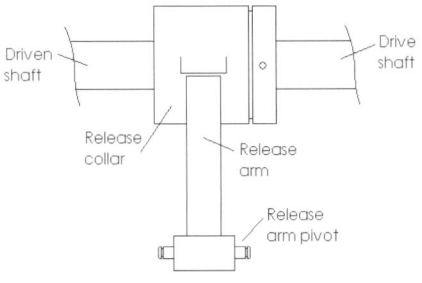

Fig.142 shows the normal mechanical layout of this type of clutch.

Plate clutch

This consists of a plate fixed to either the drive or the driven shaft and a floating plate on the opposite shaft. Both plates are made of iron or steel and have matching faces covered with a friction material. Both plates are partially enclosed within a coil. When power is applied to the coil the two plates are attracted to each other and the drive is via the friction coating.

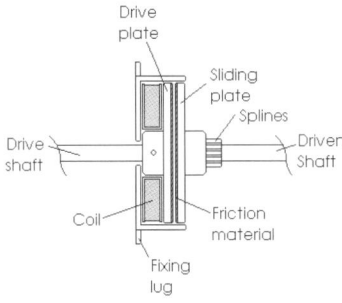

Fig.143 shows a typical electromagnetic clutch. Splines are shown as the method of flexibility of movement for the sliding plate but any form of drive that allows end ways movement whilst locking the shaft together for rotational movement would be suitable.

Serrated plate clutch

This is essentially the same as the plate clutch. Instead of friction material a series of radial serrations or teeth are on the plates.
A simpler but less compact design uses a solenoid and a drive arm to move the plates together.

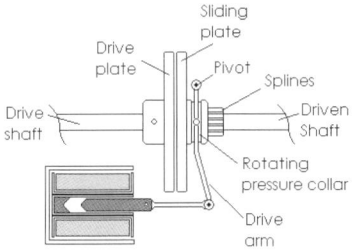

Fig.144 shows a typical arm driven clutch. When the solenoid is activated the drive arm via the rotating collar moves the sliding plate against the drive plate. Because the sliding plate has splines that are moving in splines on the driven shaft the driven shaft is pulled around with the plate. To maintain a constant pressure from the solenoid the solenoid is held just away from 'bottoming' or a strong spring can be fitted in the solenoid link arm and set such that the spring is slightly stretched when the solenoid is fully home. This can be used with a friction plate but would require a strong solenoid to prevent clutch slip.

Dog clutch

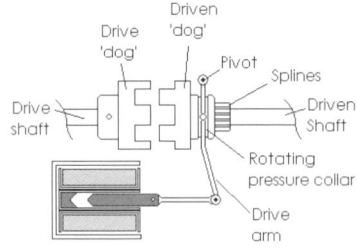

Fig.145 shows a dog clutch. This is essentially the same as the serrated plate clutch. Instead of the series of radial serrations full 'dog' teeth are

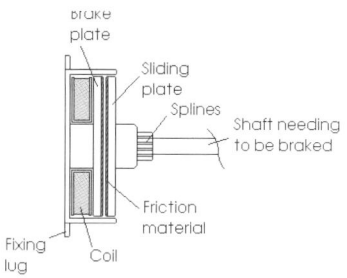

Fig.146

on the drive and sliding plates.

As with the serrated plate clutch a simpler but less compact design uses a solenoid and a drive arm to move the 'dog' teeth together.

The methods of operation are as the serrated plate clutch but because the 'dog' teeth provide a clutch that is less likely to pull out of mesh accidentally a smaller solenoid can be used. It will probably be necessary depending on the loading to provide a 'lead in' to the tooth profiles.

Brakes

This is essentially the same as the plate clutch. But uses a floating plate on the shaft that needs to be braked and the other plate fixed to the frame of the unit. When power is applied to the coil the moving plate is attracted to the fixed plate. The friction coating stops or slows down the rotation.

Fig.146 shows a typical electromagnetic brake.

CHAPTER NINE

Interference Suppresion

Interference takes a number of forms.
Electric fields are generated when contacts break and arcing occurs. This occurs for example at switches and the commutator of a motor. This kind of interference can travel relatively long distances and creating sparks was the basis of early experiments for radio transmission.
Magnetic fields are produced when currents flow in a wire. The higher the current in the wire the greater is the magnetic field associated with that wire. If this current is AC or changing quickly in for example PWM driver circuits, this field can be induced into other wires in close proximity.
Conducted interference can be carried along power supply wires or along signal cables. Interference suppression takes two main forms these are prevention and blocking.

Mains input suppression

Input suppression is easily achieved by the use of an inlet filter. These can be separate units or can be obtained as part of the inlet socket. They prevent any mains borne interference passing into the unit and hopefully any interference produced in the unit passing onto the mains. These units can be built from the special X and Y type capacitors but because of the potential it is safer to use a ready-made unit specifically

Two small motors showing input mounted suppression boards.

Ingress Protection Rating (IP)					
Solids – first digit		Fluids – second digit		Impact – third digit	
Level of protection		Level of protection		Level of protection	
0	None	0	None	0	None
1	Solid objects over 50 mm	1	Vertical falling drops of water	1	Impact of 0.225 joule – 150gm weight dropped from 15cm
2	Solid objects over 12mm	2	Direct spray up to 15° from vertical	2	Impact of 0.375 joule – 250gm weight dropped from 15cm
3	Solid objects over 2.5mm	3	Direct spray up to 60° from vertical	3	Impact of 0.5 joule – 250gm weight dropped from 20cm
4	Solid objects over 1mm	4	Water spray from all directions – limited ingress	4	No meaning
5	Dust – limited no harmful deposits	5	Low pressure spray from all directions – limited ingress	5	Impact of 2 joule – 500gm weight dropped from 40cm
6	Dust – total protection	6	Strong pressure spray from all directions – limited ingress	6	No meaning
7	No meaning	7	Immersion between 15cm and 1m	7	Impact of 6 joule – 1.5kg weight dropped from 40cm
8	No meaning	8	Long periods of immersion – under pressure	8	No meaning
9	No meaning	9	No meaning	9	Impact of 20 joule – 5kg weight dropped from 40cm

designed for the task.

Transient spikes can still be a problem and transient suppressors are available for connection across the mains. These components do nothing under normal working conditions but if a spike above the rated voltage occurs the impedance of the device changes quickly effectively clipping the spike.

Enclosures

There are many methods of preventing airborne interference and this can go from totally enclosing metal boxes, copper beryllium finger strips to ensure that removable panels are in contact with the main section of frame, through special connectors and compatible fan and filter covers.

Metal boxes will prevent electric fields if the box is connected to ground. Magnetic fields are only stopped by iron or to a slightly lesser degree by steel boxes. Aluminium is transparent to magnetic fields.

Holes in enclosures will allow electric fields to pass. The frequency of the electric field that passes through is relative to the size of the hole. Enclosures are also rated in most European countries under the ingress protection rating (IR) laid out in IEC529 (BS EN 60529 1992). This is not directly related to interference but can give some guide to the holes in the enclosure that can be a potential source of interference either in or out. The rating is a three-digit number the first digit relates to solids, the second to fluids and the third to impact.

Cable runs

Wires can pick up spikes by being in close proximity to an interference source. This source can be another wire.

Neat careful routing and keeping high current switching wires away from logic cables usually suffices.

Using wires as twisted pairs helps to cancel out any interference pick up. If all else fails then it may be necessary to use screened cable runs but it is important that only one end of the screen is attached to ground otherwise an earth loop is formed.

Capacitors can be added between the end of a wire and ground to remove spikes but may have a tendency to slow down logic signals slightly. Low voltage transient suppressors can be used on cables either producing or susceptible to transient spikes.

It was common practice to fit zener diodes on wires for spike suppression. This was very common with long run wires such as those on communications equipment.

Earthing

Earthing or grounding is necessary in most equipment as a first stage. It is important that earthing cables are of as low a resistance as practicably possible.

All wires and cables have a finite resistance. There can therefore be a difference in potential across wires connected to different ground points. To prevent this, where possible all earth wires should be taken to a common earth point. This prevents the creation of an earth loop. In practical terms running a wire back to an earth point unless the wire is low resistance and by implication large cross sectional area may create more resistance than using the chassis as a return point.

Screened cables should have the screen connected to ground but only at one end.

Capacitors

Large electrolytic capacitors are commonly used for smoothing in DC power supplies. By virtue of the construction an electrolytic capacitor it normally exhibits a high inductance. In most circuits where the frequency is low such as 50/60Hz mains power supplies this is of no consequence in practical terms. Where high frequency noise is concerned the effect will be similar to the current rise in a motor winding and the capacitor will 'block' the high frequency and it will not be suppressed. Adding a second low value capacitor in parallel with the main capacitor will allow the suppression of the high frequency.

Small capacitors should also be attached across the supply to each IC as close as possible to the IC for the same purpose and at the inputs and outputs of regulators. A good starting point for interference capacitors is 100nF or 0.1µF.

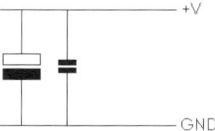

Fig.147 shows a typical parallel capacitor pair.

Motor spikes

With the circuits in this book the greatest potential for interference problems is from the DC motors. This can normally be suppressed by the use of capacitors or a combination of capacitors and chokes or ferrite beads.

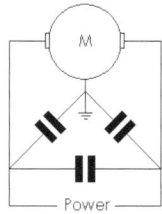

Fig.148 shows a simple form of capacitor suppression. The earth point is normally the motor body and is not usually returned to ground via a wire although the motor mount may provide an earth path.

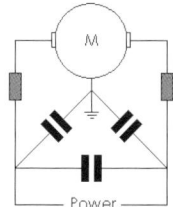

Fig.149

Fig.149 shows capacitor and ferrite bead suppression. The ferrite beads are threaded over the connecting wire. If the wire is put through the wire once the bead hole should be as close as possible to the diameter of the wire. Passing the connecting wire through the bead a number of times can increase the inductance.

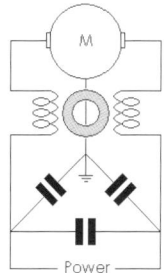

Fig.150 shows capacitor and ferrite differential choke suppression. The windings on the ferrite ring choke are wound in the out of phase differential mode. Differential mode acts as common mode noise cancelling.
 The best method of mounting this type of suppression on a motor with push on connectors is on a small PCB the same diameter as the motor soldered directly to the push on connectors.

Electro-magnetic compatibility directives

This is a collection of rules covering allowable 'interference' levels. Most countries have an equivalent e.g. FCC rules in the USA.
These are aimed mainly at industry and are constantly being upgraded and it is feasible that anyone who builds a unit and subsequently sells it even if not for a profit could come under the legislation.

CHAPTER TEN

Heatsinks

Heat travels along a temperature gradient. Imagine a hill, the steeper the hill the faster you could potentially travel down the slope. The same effect occurs with heat the higher the difference in temperature the more heat will 'flow'. This is why central heating radiators run at a much higher temperature than the desired room temperature.

Whenever current is carried through a conductor or a semiconductor one of the by products is heat. With conductors a rise in temperature causes an increase in resistance. With most semiconductors the opposite is true. If a means of removing the heat from a power component carrying large currents is not used then destruction of the device is the inevitable effect. It is easy to fit a temperature measuring device that can close the system down if temperatures do approach critical.

Metal heatsinks

These are probably the most common means of removing heat from a component. They consist of in their simplest form a metal clip that attaches to

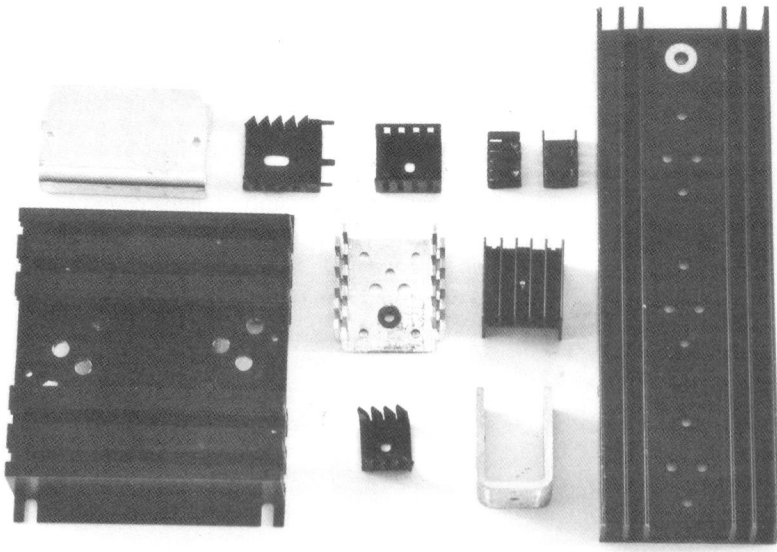

A range of heat sinks.

A heat sink assembly with three MOSFETs. The MOSFETs are PWM control, brake and 'freewheeling diode'. The spare hole accommodates the voltage regulator.

the component, or even an un-etched section of a printed circuit board, to elaborate extrusions. They all share the same way of categorising and that is in terms of °C/watt. This means that for every watt of power transferred to the heatsink the temperature will increase by a specific temperature. Small heatsinks can be rated in the order of 60°C/watt and large ones in the order of 1.8°C/watt. Consider the following examples

The HUF75337P3 that many of the MOSFET circuits in the book are based on has the following characteristics.

   Working temperature -55°C to +175°C
   On resistance 0.014W
   Maximum voltage 55V
   Maximum current 62A
   Maximum dissipation 115W

If we wish to carry a current of 50A
The voltage drop across the device would be $V = IR$
That is $V = 50 \times 0.014 \, \backslash V = 0.7$
From $P = IV$ $\therefore P = 50 \times 0.7$ $\therefore P = 35W$
If the ambient temperature **inside** the case is 35°C when the unit is operating and we do not wish the driver to exceed 80°C
The temperature difference is $80 - 35 = 45°C$
This equates to a heatsink rating of 45/35 = 1.3°C/watt
This is a large heatsink.
If we were prepared for the component temperature to rise to 150°C the temperature difference would be $150 - 35 = 115°C$
This equates to a heatsink rating of 115/35 = 3.2°C/watt
This is a much smaller heatsink.
If we only required the component to carry 10A
The voltage drop across the device would be $V = IR$
That is $V = 10 \times 0.014 \, \backslash V = 0.14$
From $P = IV$ $\therefore P = 10 \times 0.14$ $\therefore P = 1.4W$
By the same arguments as previously this is 32°C/watt for a rise of 45°C and 82°C/watt for a rise of 115°C.
These are equivalent to small clip on type heatsinks.
Heatsinks that are painted matt black are better transferors of heat than the same heatsink in the normally polished aluminium.

Forced air

This has a major effect on cooling but the results are not always fully predictable. Fitting a temperature-controlled fan can mean that heatsinks can be reduced and this can often mean smaller enclosures.

Fans come mainly in two designs for fitting into electronic units. The first is the common blade type and the second is the impeller type. There is a wide range of small low voltage DC fans that were initially made for computer installation that are suitable for electronic enclosures.

A few years ago I worked on a piece of equipment that had worked successfully in the original form for many years. A new variant was brought out but used most of the same parts in the processor and power supply units. Unfortunately within a period of months the failures of the new units in percentage terms of the installed base exceeded the failures of the older units. Virtually the only change between the two variants in the section that was failing was the mounting of the fan. On the new model the fan blew outwards instead of inwards and because of this the dust filter had been moved to a different section of the cover plate. The effect was that air no longer went through the filters it took the easier route of travelling through any gap in the machine covers. The dust was no longer filtered out but was distributed across the processor boards and power supply. This caused a blanket effect on components and caused them to run hotter because they could not dissipate heat as easily. Coupled with the fact that when air is less dense it carries less heat because the air molecules are fewer. In this case because the air was being 'sucked' out rather than being forced in less heat was being removed. The overall effect was a rise in temperature of the components and premature failure. Reversing the fans and fitting the filters has they had been mounted previously solved the problem.

The secondary effect of denser air is the tendency to circulate around components better.

Water radiator

This is an item that has progressed in fits and starts over the years. Originally automobile engines were air cooled now most are water-cooled. This theme has carried on through other engine types such as light aircraft and motorcycles. The advent of the super computers saw water cooling being used and now water-cooling is available for the home PC.

With the components used in this book I think I would initially use multiple drivers or use air cooling and larger heatsinks. No doubt water-cooling will find a place in some designs.

Peltier effect devices

This is a very interesting device with no moving parts. They do not produce heat but are able to move heat between the two faces of the device when a DC voltage is applied. The hot side and the cold side depend on the direction of the current flow. Temperature differentials of up to 70°C are possible.

With a heatsink fitted to the 'hot' side to dissipate the high temperature and the 'cold' side fitted to the component heatsink a large amount of heat can be moved.

Maxwell's Demon

This is a phenomenon that I believe is still being worked on in a number of universities and research institutes.

In 1871 the physicist James Clerk Maxwell proposed a thought experiment. A wall separates two compartments filled with a gas. The wall has small trapdoor opened and closed by his demon. Maxwell originally called it a finite being but by the 1920s the demon description had stuck. The demon allows all molecules with higher than average energy to travel into one compartment and all those with lower than average energy to move into the other compartment. Eventually one compartment would be full of high-energy particles i.e. hot and the other compartment would be full of low energy particles i.e. cold.

There were arguments about whether this was against the second law of thermodynamics. It is not but is in fact simply a redistribution of kinetic energy. As a child I had an engineering book that tried to explain complex scientific processes in simple terms. This book was fascinating for me as a child and probably would be even now. It contained projects such as how to build your own pulse jet engine from a piece of metal tube. Unfortunately the book disappeared over the years and I have never come across another copy. I do not know the name of the author so I am unable to give him the credit. One of the projects was titled something like 'Taming Maxwell's Demon'. The author had produced a device and provided a number of tables referencing variations of the design to temperature difference and airflow rate. The unit was simply a tube with a shaped section in the middle. High-pressure air pumped in through the side of the tube created a vortex. High-energy air particles would travel around the outside of the airflow probably collecting even more heat from the pipe by friction. The lower energy particles fell to the centre of the airflow. A baffle that was basically a washer separated the two flows and the effect was that hot air came out of one end of the tube and cold air came out of the other end. This was a simple example of the redistribution of energy levels.

Cold air can remove more heat than hot air and this device may find use in localised cooling.

Heatsink mounting

The mounting tabs of some power component are not isolated from one of the terminals. This means that an insulator needs to be fitted between the component and the heatsink. These insulators may be mica or various forms of plastic. Some types may also require thermally conductive grease.

A neater way of mounting a large number of power components is with thermally conductive adhesive. This produces a thermally conductive but electrically insulating bond. A strip clip can be added to hold the component while the adhesive sets or can be left permanently in place.

CHAPTER ELEVEN

Fuses & Circuit Breakers

Fuses and circuit breakers take two general forms, devices that protect the machine and devices that protect the operator. An example of the first category is fuses and an example of the second is the RCD – residual current device. With the exception of the mains driven power supplies all the circuits are low voltage. Therefore this section of the book is only dealing with devices to protect the machine. General safety issues are dealt with in the safety section.

Standard fuses

The most common form of these consist of a glass or ceramic tube with metal end caps that clip into a holder. The tube carries a wire that joins the metal end caps. The wires are of a size that will melt if the current rating is exceeded by a specific amount for a specific time. Fuses vary greatly in their characteristics therefore data sheets should be checked to find a suitable

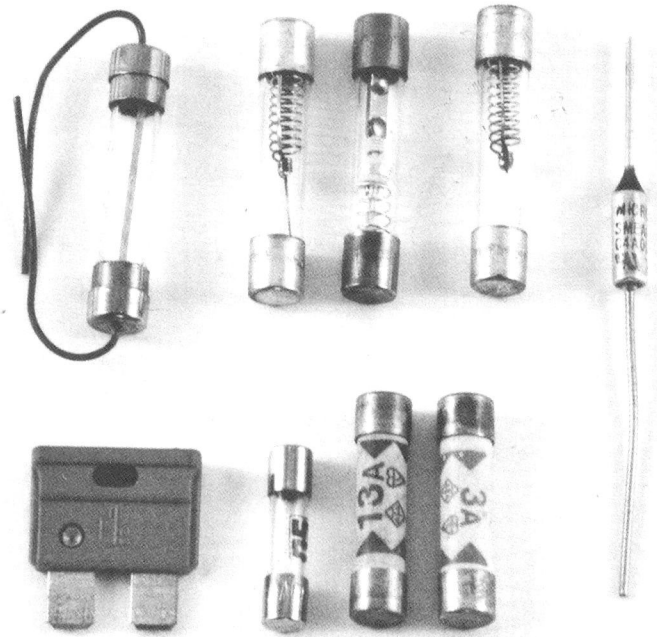

A range of fuses. The metal cased fuse on the right is thermally operated.

A self re-setting electronic fuse.

device. There are also a large number of variants of body style such as blade fitting and wire ended but they all work in the same manner and are of once only use.

A variant of the standard fuse is the slow blow fuse that is designed to stand a surge current. These are used in applications such as motor circuits where the start current may be many times the run current. When they are in a glass packaging they can usually be recognised by the spring attached to one end of fusible link.

Electronic fuses

These should be really called auto-reset circuit breakers. They consist of an electronic circuit that will stand a large over current but will trip immediately on a short circuit. After a period of time they will reset but if the fault condition still exists they will immediately trip again.

These devices are usually available with values between 1 and 10A.

It is essential that auto reset circuit breakers should not be relied on to protect any machine that could restart when the device resets. If they were used on a machine that for instance 'tripped out' because of an overload and the overload was removed the machine would start off without warning when the device reset. It may be said

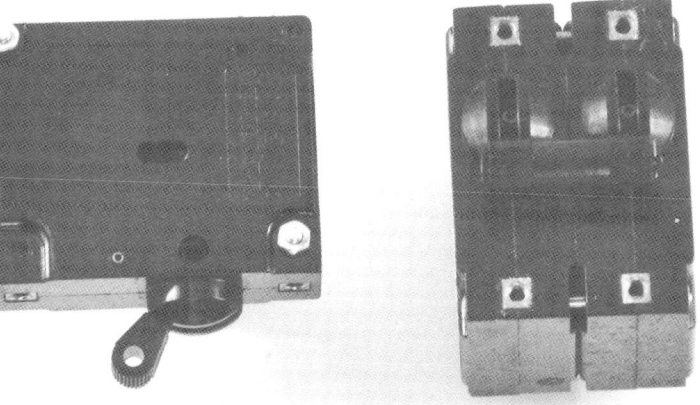

This shows a double and single low-voltage MCB both available in a wide range.

Fuses & Circuit Breakers

that the applications in this book are only for small devices but with the use of gearing and the speeds and power some of the small DC motors can achieve then the consequences could be catastrophic. These devices should only be used as part of a total cut out or NVR system. This is discussed further in the safety section.

Thermal fuses

These are not re-settable. They consist of a heat conductive tube surrounding a wire of known melting point. They are placed in close proximity and preferably in contact with the part being monitored. They act as a fuse but are activated by the temperature not the current flow.

Re-settable fuses and MCBs

These devices are manually reset and take two forms. The oldest type consists of a heating wire wrapped around a bimetal strip. Current flow causes the wire to become warm and the bimetal strip bends unlatching the contacts. This principle is also used in AC motor contactors. The heating effect takes time to bend the bimetal strip so the device can stand high current surges but trips on a predicable continuous overload. The second device is the low voltage MCB – miniature circuit breaker. This works on the same principle as the high voltage circuit breakers found in the majority of 'fuse boxes'. A coil of wire fits around a core that attracts an arm attached to the latch mechanism. When the current is at a certain level the magnet force will be sufficiently strong to unlatch the contacts. Both these type of devices need to be manually reset and are therefore inherently safer than the auto reset type. Before resetting it is imperative that the condition that caused the initial problem is resolved and that no other danger ensues when the device is reset.

CHAPTER TWELVE

Inputs

Most control circuits require some form of input. This can vary from a simple mechanical switch to complex electronic sensors.

Mechanical switches

Switches in basic form are simply mechanical devices that move a contact from the normal at rest position to a pressed position. The switch may be momentary i.e. moves back to the rest position when released or may be stable in both states i.e. it will remain in one position until physically moved to the other position.

On/off

Push to make
These are switches where one or a number of contacts close when the switch is pressed.

Push to break
These are switches where one or a number of contacts open when the switch is pressed.

Switches in electronic circuits

Electronic circuits require only small input currents but this signal may need to be in the

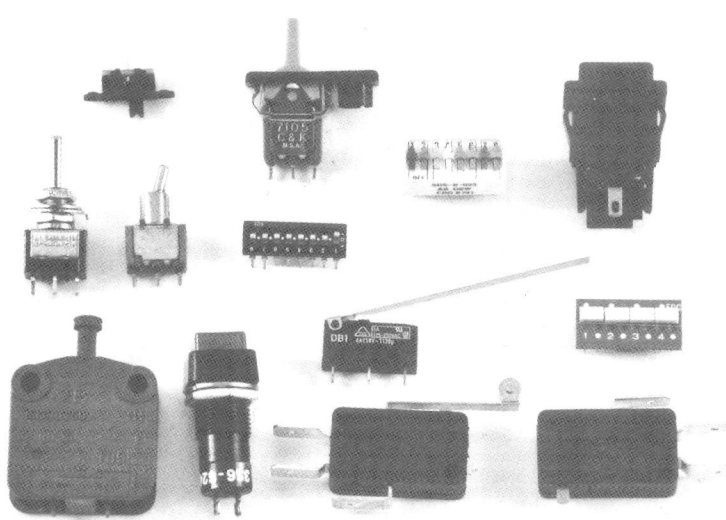

A range of micro, slide and push switches.

Switch position	O/P 1	O/P 2
Centre	High	High
Up	Low	High
Down	High	Low

TABLE A

Switch position	O/P 1	O/P 2
Centre	Low	Low
Up	High	Low
Down	Low	High

TABLE B

form of a 'pull up' or 'pull down' input. In the case of a push to make switch, pull up is when the input pin is tied to the negative line with a resistor; the switch is between the pin and the positive rail. When the switch is closed a positive input is placed on the pin. Pull down in this case is with the resistor tied between the pin and the positive rail and the switch between the pin and the negative rail. The input pin is normally held positive, it is pulled negative when the switch is pressed. The inverse would apply with push to break switches. High current can be controlled from switches capable of switching only a few milliamps using relays or electronic driver circuits.

Fig.151 shows the possible options for a single switch.

a) Produces a positive to negative transition with the press of a normally open switch
b) Produces a negative to positive transition with the press of a normally open switch
c) Produces a negative to positive transition with the press of a normally closed switch
d) Produces a positive to negative transition with the press of a normally closed switch

Changeover switches

These are a single or multiple of a push to make and a push to break switch in one body but sharing a common centre contact. These again can be obtained in momentary or stable form. There is also a version with a centre off, these are often momentary with the centre position being the stable state but other options are available. The centre position is a both contact off position.

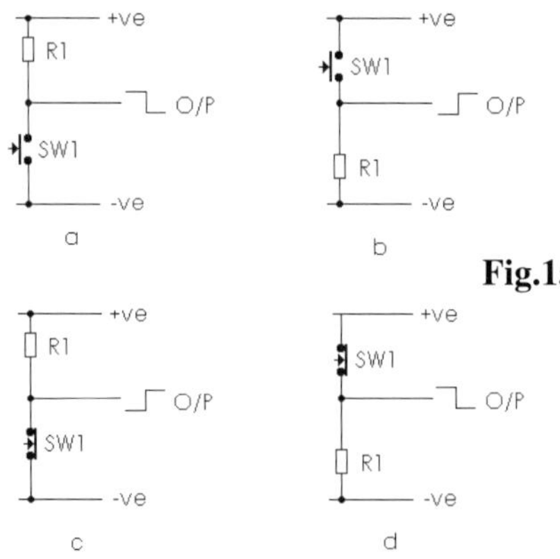

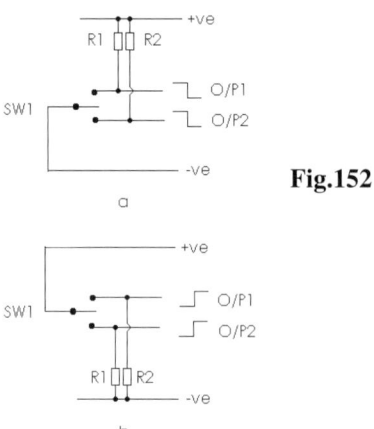

Fig.151

Fig.152

108 Electromechanical Building Blocks

Fig.152 shows a centre off changeover switch giving three outputs that can be set up to give various options.
a) Shows an active low version – table a gives the outputs
b) Shows an active high version – table b gives the outputs

Switch bounce

When a mechanical switch closes it does not close cleanly but can bounce for up to 0.1 seconds. This can be seen by fast moving logic as multiple switching. There are a number of electronic switch bounce circuits but the easiest means is probably to use a small capacitor by itself or in combination with a Schmitt trigger circuit. The Schmitt trigger is a device with switching levels that are above the normal logic switching level and ensures that switch on and switch off occur quickly when the trigger point is reached. Using a fast acting comparator circuit can simulate an adjustable level Schmitt trigger. The effect is that the voltage rises exponentially across the capacitor and therefore smoothes out the switch bounce.

With many of the circuits in the book switch bounce will have minimal effect because it does not mater in reality how many times you switch something on it is still on. Problems really only occur in circuits that count or multiple events occur on subsequent switch operations.

Fig.153 shows a capacitor being used for switch bounce suppression. The connection of the switch and capacitor will depend on whether the switch is needed to produce a positive or negative going output.

Fig.154 shows a capacitor being used for switch bounce suppression in conjunction with a Schmitt trigger.

Fig.155 shows a monostable being used for switch bounce suppression. A retriggerable monostable is used where the length of the monostable pulse must exceed the small low level bounces at switch on and switch off. Once triggered the monostable will remain on during bounces and whilst the switch is held. The monostable time must exceed the likely 'gap' between bounce pulses otherwise two counts will be registered.

Multi pole

These are switches with multiple output configurations. They are easily simulated from simple switch inputs but using simple logic four lines can easily be converted to sixteen single outputs.

Rotary

These are switches that as the name implies move in a rotary direction. They are normally limited to a maximum of twelve positions. Complex switches are often made up from switch wafers to build up the switching operations. They are expensive both as a switch and in the time taken for wiring complex arrangements. The current switching capability is often low. There are switches available that appear to be rotary but are really decoded sequential switches.

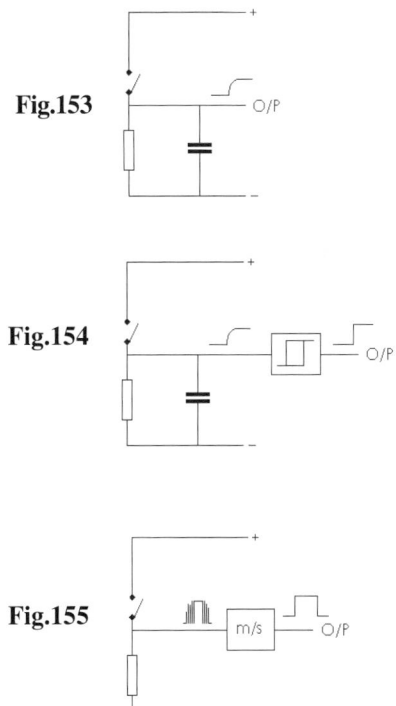

Fig.153

Fig.154

Fig.155

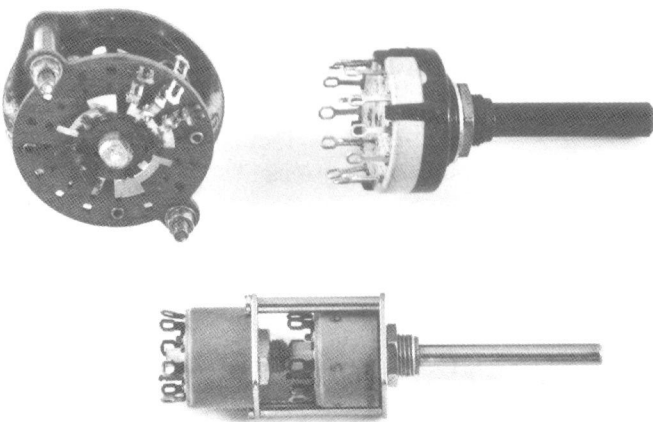

Typical rotary switches.

Rotary switches can be simulated by other means and an easily made rotary switch can consist of a disc with a magnet attached moving past a number of reed switches. Some form of detent is required to hold the magnet at each reed position but this can be as simple as a number of grooves and a sprung roller. The output can be decoded using simple logic to simulate many types of output.

Binary and BCD

These produce an output using a common and four-switched line in the form of 0 to 9 for BCD, and 0 to F for Hex. These can be used with diode 'OR' gates to produce multiple outputs from one position.

Gray code discs

The Gray code is a coding that changes only one bit per step. There is less possibility of wrong outputs due to transitions. With standard binary a number of bits change. On a circular binary 8-bit encoder moving from 255 to 0, there are eight bits changing at the same time. A Gray

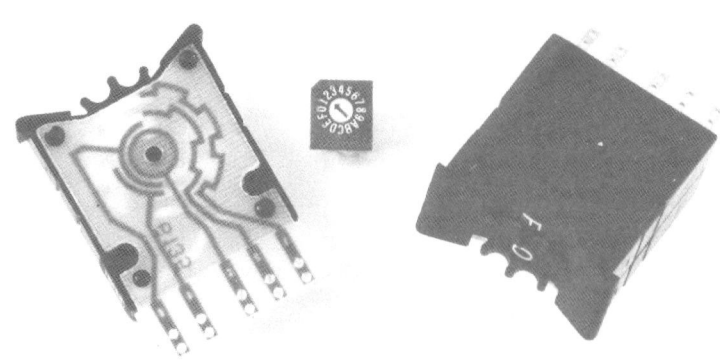

BCD output rotary switches. The larger type is finger operated whilst the smaller has a screwdriver slot.

code disc is usually in the form of an encoder instead of a switch. The difference is that the encoder does not have detents, the disc is free to rotate under the influence of some external stimulus. An example of a Gray code encoder is in a wind direction or compass readout. If this required to a resolution of one degree, the device would need a 9-bit output. This is a very expensive encoder disc and unit to produce. See earlier section for in depth explanation of Gray to binary and binary to Gray encoding. If for some reason the extra conversion between a Gray disc and binary is not practical then it is possible to use a binary disc with strobe holes. The strobe holes consist of a complete track of holes, one for each read position. The strobe holes are only half the width of the normal holes in the disc. This extra strobe output will only occur when the other holes are centralised over the sensors therefore the likelihood of binary output errors is minimised. The strobe output can be used to trigger a latch or enable a series of and type logic gates.

Special coding discs and strips

Most discs use a standard coding in a pattern of zero to maximum bit value. There are occasions when something different may be required. There is no reason why any coding sequence needs to be in disc form it can as easily be laid out as a strip or a quadrant or any other practical shape. It is also possible to set out any sequence of a bit pattern. Consider a quadrant on a joystick motor speed controller. The centre position could be zero with braking. The minimum PWM that the motor will run at in the application may be 25% therefore the first bit code would be equivalent to this value. The value could then increment with movement to a maximum value. The hole pattern in the quadrant could be duplicated the other side of the zero with one extra bit in the pattern to indicate reverse. With the appropriate decode logic there is a joystick that would move from zero to a maximum value but would brake the motor in the zero position before moving in to reverse.

Sequential binary code discs

Many types of counting discs move in one direction only or the direction is not relevant to the count e.g. speed measurement discs. Where it is necessary to also measure direction of travel or to produce an output that is relevant to the distance moved and the direction different techniques need to be applied.

Sequential binary code discs can be used for up/down counting where the timing disc can move in either direction as part of its normal operation. Using a disc with 180 slots plus a home marker provides full 360 degree readout. This is achieved by making the disk pass through the 0 (home) position after switch on. Up count/ down count sequences can be processed and moved to a counter. Using edge transition counting i.e. counting 0 to 1 and 1 to 0, the 180 slots are doubled to 360. A disc with 180 slots is easily within the capability of laser printers and most inkjet printers. The resulting disk print can be etched through a thin metal such as brass or stainless steel or used printed onto film and used with an optical sensor.

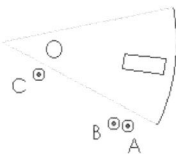

Fig.156 shows a section of a circular timing disc. Only one hole is shown. A and B are offset sensors. If an output from sensor A is followed by an output from sensor B the disc is moving clockwise. If an output from sensor B is followed by an output from sensor A the disc is moving counter clockwise. Sensor C is the home position sensor. This is used to keep the count in synchronisation. There is only one hole in the disc to represent home position. Problems occur with this simple set up if the disc stops exactly over the sensors and then moves back in the direction it was travelling from. Extra logic can be

used to counteract this effect this is explained in more detail later in the book.

Ladder networks

These are used to turn digital inputs into analogue levels. Discrete components can be used or a few dedicated ICs are available. Most use R – 2R resistor networks or simpler forms use R - 2R - 4R resistor networks.

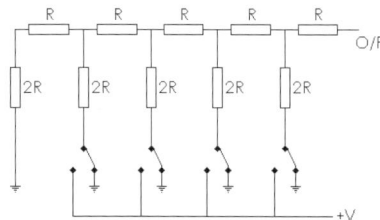

Fig.157 shows a 4 bit R – 2R network. The output voltage is dependent on whether the 2R resistors are connected to ground or to the positive rail. The relative output voltage will depend on the input voltage.

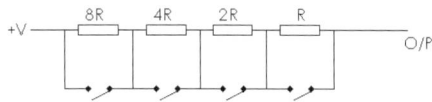

Fig.158 shows the simpler R – 2R – 4R network. The output voltage depends on the input voltage and on the number of resistors shorted out. The circuit is easier to build but requires resistors to be made up individually.
All electronic switches have a finite resistance. With the R – 2R network this can be accounted for as part of the 2R resistance. With the R – 2R – 4R network the resistance of the electronic switch is a different percentage of each ladder step. This can lead to a small non-linearity of the output depending on the switch resistance compared to the minimum and maximum resistance.

Reed switches

These are normally open contacts or changeover contacts in a sealed glass tube. A magnet in the correct orientation to the glass body activates the switch. They are useful in environments where a normal switch may have problems i.e. dusty or damp conditions. The reed switch does not need direct contact with the magnet. The switch distance will depend on the reed characteristics and the strength of the magnetic field. A relay version is available which uses a coil to produce the magnetic field. These are usually capable of being driven directly by logic level outputs.

Latching or 'first on'

Latching switches are often met on domestic items such as radio frequency selection. They consist of a switch, usually multi-pole and a mechanical latch. When a switch is pressed it latches and releases the latch on any other switches pressed. It is easy to press and latch down more than one switch on most designs. This type of switch is not suitable for critical areas. A similar operation can be achieved using discrete switches and a large amount of logic, but even these are not always foolproof because some designs allow an output, albeit not latching, if multiple switches are held down.

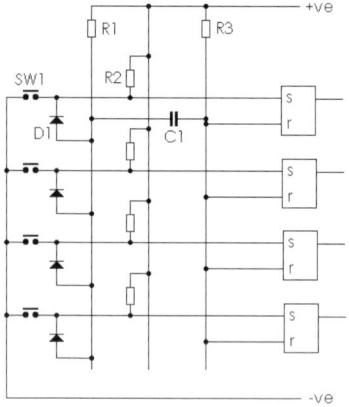

Fig.159 shows a simple selection circuit that produces a latched output when the appropriate switch is pressed. This diagram uses negative switched latches, but a similar circuit can be

A range of optical sensors and a LDR.

made using positive switched latches and using the negative rail and reversed diodes.

When SW1 is pressed, the set input of the first latch is taken negative causing the output to be set. At the same time D1 pulls the R1 line to ground causing a negative pulse to appear on all the reset lines via C1. This will reset any other previously set latches. The same effect occurs with any of the other switch and latch sets. The circuit can be extended to any reasonable number of switches and latches.

Problems occur if one switch is held down and another is pressed. Two outputs can latch because the reset is only a pulse. This can have serious effects on the next driven stages of the circuit. It is feasible to add interconnecting logic to the circuit to prevent this but the more stages that are added; the more complicated and unwieldy the circuitry becomes. Another problem is pressing the switch quickly, this will cause a set pulse that is shorter than the reset pulse, and therefore no outputs will be latched.

Sequential switching

These take three forms. The first is similar to the up/down count described in the Gray code encoder. The second is similar but use two pairs of up and two down count switches. They are often arranged in the form of a joystick. An output which can be in form of numbers or pictorial rows and columns are moved over until an output is selected and another switch is pressed for the event to take place. Simple logic will handle this type of switching easily. Extra features such as single step or multiple steps after holding for a period of time, similar to a computer keyboard, can be added. The third form is using an optical disc with offset sensors similar to the binary sequential disc. The order of the sensors switching determines both the count direction. Single step or multiple steps after holding for a period of time, similar to a computer keyboard, can be added to this type of circuit. Problems can occur with this type of switch when using conventional logic. If the disc stops exactly between sensors and then moves in the opposite direction, the count sequence can be lost. Logic gates can be added to prevent this.

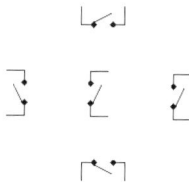

Fig.160 shows the typical layout of a joystick type switch. Neither up and down or left and right can be pressed simultaneously due to the physical layout.

Optical

These switches usually contain an LED and a sensor in the form of either a photodiode or a phototransistor. There are high gain versions with built in amplification. Some versions are available as reflective where the LED and sensor are in the end of the unit and angled so that the light from the LED is reflected back to the sensor when a reflective section passes the focus point of the LED and sensor. Most optical switches use transmissive mode where an LED shines directly at a sensor behind a narrow slot. A vane

Inputs 113

or toothed disc is used to block the light. This type of switch is very good for position sensing or event counting but is unreliable in dusty environments. A few of these optical switch outputs swing very close to the power rails so that they can be treated as low power switches. Many are better fed through an OP amp or comparator to obtain the switching levels.

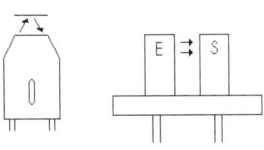

the opto sensor. The output transistor is set up with the emitter tied to ground i.e. it can sink current but it cannot source it. A resistor is used to supply current and the output will be positive when the phototransistor is not conducting and near ground when the phototransistor is conducting.

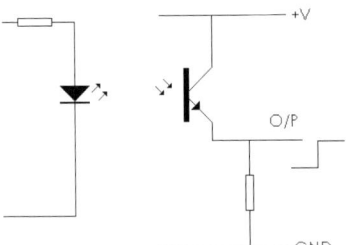

Fig.161 shows the typical case style of a reflective and a transmissive opto sensor. The focus point of the reflective type is typically in the order of 4mm.

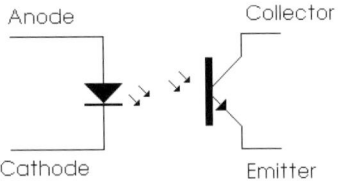

Fig.162. This diagram shows the typical internal circuit of a phototransistor output for both the reflective and transmissive types.

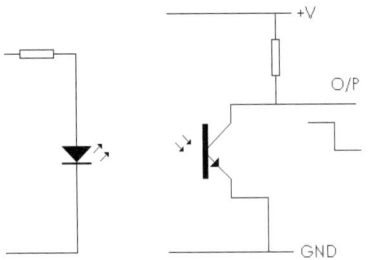

Fig.163 shows the one method of connecting

Fig.164 shows another method of connecting the opto sensor. The output transistor is set up with the collector tied to positive i.e. it can source current but it cannot sink it. A resistor is used to sink current and the output will be near ground when the phototransistor is not conducting and positive when the phototransistor is conducting.

Hall effect

There are two types of Hall effect sensor. The first type produces a linear output that varies with the strength of the magnetic field. The second type more commonly used for position sensing is the type that gives an output at a specific magnetic field strength. These may have signal conditioning built in.

Some types have a very small voltage change in the presence of a magnetic field. This change may be as small as 0.1V. The actual change and the direction of change will depend on the magnetic field strength and the polarity. The output is usually fed to an OP amp or comparator to obtain the switching levels. With the common types of sensor a south pole on the face or a north pole on the back produces a rise in output. Most three-leg sensors have markings on the front of the sensor. The active area is small compared to the sensor face and is usually on the vertical and horizontal centre line.

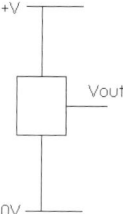

Fig.165 shows a three-lead device connection. Some of these devices have a frequency response up to 100kHz.

Fig.166 shows the standard set up for a Hall effect sensor. A moving magnet or a number of magnets in succession can be brought close to the sensor or a fixed magnet can be interrupted by a steel or iron vane. The vane can be in the form of sectors cut from a disc.

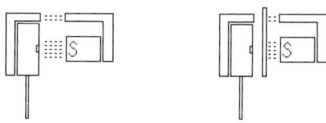

Fig.167 shows an arrangement for a concentrator. The concentrator allows the use of less powerful magnets or greater separation. Direct magnetic influence or a vane can still be used.

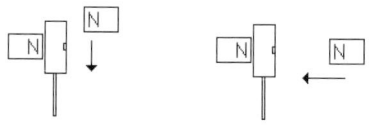

Fig.168 shows the sensor biased on by a weak magnet at the back of the sensor. A stronger magnet in proximity to the face will turn the sensor off. Direct magnetic influence or a vane can still be used.

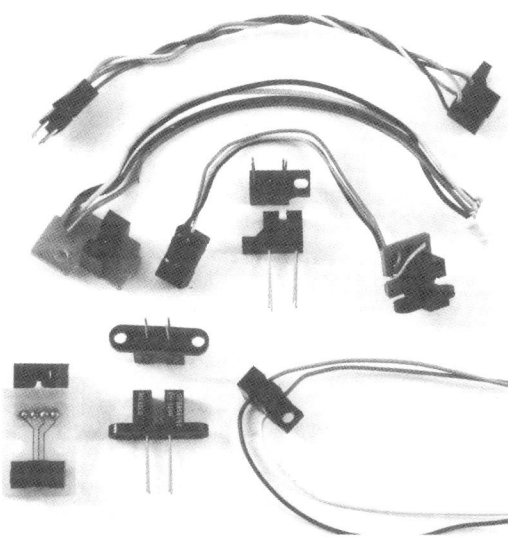

A range of optical disc/vane sensors.

Fig.169 shows the sensor as a gear tooth sensor.

With the north pole at the back of the sensor absence of ferrous metal is sensed. With the south pole at the back of the sensor presence of ferrous metal is sensed. The sensor will normally need to be very close to the gear teeth.

Fig.170 shows a gear tooth sensor where a ferrous metal concentrator is used. Use of a concentrator produces a greater field change and allows greater separation of the sensor and gear teeth.

Magnetoresistive sensor

This sensor is similar to the Hall effect sensor in that it reacts to magnetic fields. The magnetoresistive sensor as the name implies, a

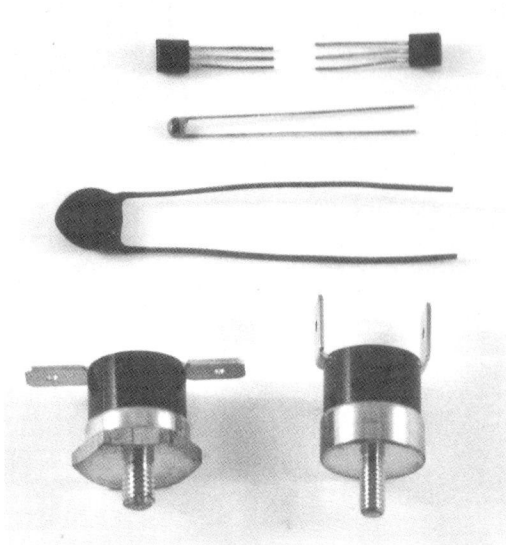

Temperature sensors showing two electronic, two temperature coefficient and two bi-metal temperature switches (thermostats).

resistance that changes with the magnet field. They are formed of metal alloys and can be produced to give a rise or a fall in resistance. They come in the form of a single resistance with two connections or a four connection Wheatstone bridge versions. In the Wheatstone bridge version two of the resistors increase in resistance and two of the resistors decrease in resistance. This produces a large output change for a small magnetic change. The magnetoresistive sensor can replace Hall effect sensors in most applications.

Current transformer

The Hall effect sensor and the magnetoresistive sensor are also available as current 'transformers'. This is in effect a metal, usually iron, or a ferrite ring with a slit into which is fitted a sensor. The current carrying wire is fed through the ring. The current produces a magnetic field in the ring. The Hall effect sensor produces a voltage proportional to the supply voltage and the magnetic field and the magnetoresistive sensor produces a change in resistance proportional to the magnetic field. Most types will work with both DC and AC current but care must be taken in choosing units referred to as current transformers because some work only on AC.

Practical circuits for high current sensing and measurement are shown in the battery current monitors and cutouts section.

Key position keyboards

These are usually laid out in a matrix formation to give 12, 16 or 20 keys. Decoder logic is needed to produce a usable output and dedicated ICs are available. Keyboards with more than 20 keys usually have custom made drivers but a few are available as standard ICs.

Sequential keyboards

These are often made using a joystick with a select button. The switches on the joystick are used to move left/right or up/down and some form of indication is used to highlight the selection. The select button is then pressed pushed to enter the selection.

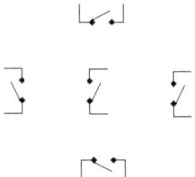

Fig.171 shows a typical layout for a sequential switch keyboard. One of the connections to each switch may be taken to a common point allowing the use of a six core connecting cable. The usual use is similar to a computer keyboard with repeat key function. A short press allows one step but holding after a time out produces a delay then multiple steps. This function can be produced using simple logic with a monostable giving a delay that can be varied and a gated astable with variable cycle time giving the repeat pulses.

Tilt switch

This is a small sealed container with a contact that opens and closes depending on the angle.

They were originally made using mercury but now substitutes are used.

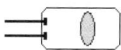

Fig.172 shows a typical form of tilt switch.
Large variable resistors or rheostats
These are almost by implication power control devices. In use one of the ends of the rheostat was attached to the supply and the slider was attached to the unit to be controlled. They worked by dropping large voltages and hence large amounts of heat were produced. Electronic power control devices have mainly superseded them.

Potential dividers

The difference between a variable resistor and a potentiometer or potential divider is that the potential divider has both ends of its resistance attached to different level voltages. The slider gives a voltage relative to the connected voltages and is acting as a voltage source rather than a resistance. Small potentiometers are used as variable resistors in some electronic circuits such as timing circuits where they are used to control the charge rate of a capacitor. Potentiometers are simple devices that can be used in many applications. They can be used for fixing a set point either directly into the ADC or via a voltage comparator. They can be used as a position indicator when attached to a mechanical drive.

Isolation techniques

The simplest isolation device is a capacitor. It is capable of blocking DC levels and allowing through AC levels. Care must be taken if the peak levels of the AC could be outside the specification of the device being driven because this high level will not be prevented unless a zener diode or some other form of clipping is used.
The shape and size of the output will depend to some extent on the size of the capacitor. A large capacitor value will allow a full waveform through as in 'A' but the height will be clipped if this exceeds the zener voltage. The rise and fall time may be an exponential shape if the frequency rate is low in comparison to the size of the capacitor. A very small capacitor will give an output as in 'B' where the capacitor charges quickly initially and the rate of change with time then falls. When the input waveform returns to zero the capacitor then discharges to produce a

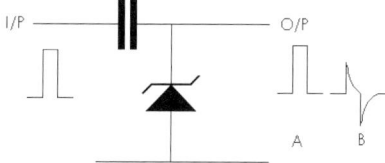

negative pulse. **Fig.173** shows a typical circuit.

Pulse transformers

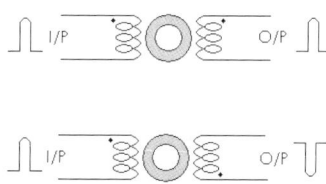

Fig.174 shows a typical pulse transformer. These are used to transfer trigger pulses to circuit devices. They are useful for triggering high voltage devices such as thyristors from low-level DC circuits whilst maintaining electrical isolation. Winding two separate coils around a ferrite core or ring easily makes these types of transformer. The same applies to peak levels as with capacitors
Changing the relative direction of the winding can change the relative direction of the pulse. The level of output pulse is determined partly by factors such as core material and input pulse

length but mainly by the winding ratio.

Optical isolation

These components are available as low voltage to low voltage and low voltage to high voltage mains AC coupling. It is common in electronic circuits for part of the circuit to have logic levels of 5V and be controlling driver circuits of 24V. It is possible that with some types of failure the higher voltage could be fed to the logic driver circuits with disastrous results. The opto isolator will prevent this happening that is why opto isolators or opto couplers as they are sometimes called are often used on PC outputs when the PC is being used to drive external circuits. The device consists of an LED driven by the logic circuit and an opto transistor or opto Darlington transistor output or the output may be processed to provide higher and faster drive capability. The output is directly dependent on the supply voltage and not on the input voltage. The low voltage to low voltage opto isolator is capable of DC or AC control and output.

The electronic connections are similar to the optical switches. There are a large number of different types and speed capabilities. In some transistor types the base may be accessible to provide biasing.

Other input sensors

It is possible to measure or provide feedback from any physical event or happening. The means are often indirect but the results are directly related to the event.

Input signal conditioning

The input of many devices can be easily driven because of the low power requirements of logic gates or driver inputs. For indirect connection in voltage level sensing mode there are many OP amps and voltage comparators that are suitable.

Temperature

Temperature sensors take two forms. The first being a fixed temperature where a mechanical switch either opens or closes at selected temperature e.g. a thermostat. The second type is an electronic IC producing an output relative to the ambient temperature.

An example of this is the LM35 that produces an output of 0.1 volts per degree Centigrade. The LM35DZ operates in the range of $0°C$ to $+100°C$ and the LM35CZ operates in the range of $-40°C$ to $+110°C$. The output of this device is linear. Another device used for temperature measurement is the thermistor. This is a device that undergoes a change in resistance for a change in temperature. These devices are available in PTC i.e. positive temperature coefficient where the resistance increases with an increase in temperature and NTC i.e. negative temperature coefficient where the resistance decreases with an increase in temperature. These devices do not produce a linear output but the repeatability is good. The resistance to temperature point can be derived from the manufacturer data usually in the form of a graph or by practical means. They work typically in the range of $-55°C$ to $+155°C$.

Another device is the thermocouple that consists or two dissimilar metals twisted together. Heat will produce a very small voltage. Outputs are typically a few tens of micro volts per degree centigrade. They can be used for temperatures up to about $1100°C$ but normally require signal conditioning interfaces that amplify and cancel noise collected by the probe. OP amps used in differential mode can be used or instrumentation amplifiers are available.

The thermostat can be read as a normal switch by the input line. The LM35 can be read as an analogue voltage by an ADC or can be fed to a voltage comparator to give the equivalent act of a switch. Change of temperature is a relatively slow event when compared with electronic speeds. The mechanical thermostat has a natural hysterisis i.e. the contact close point and contact open point is not quite the same. With an OP amp even with a feedback circuit there is little natural hysterisis. In applications such as cooling where it is desirable to switch on for example a fan when a specific temperature is reached fast cycling can occur or even oscillation where the temperature is at the transition point. True voltage comparators can be worse because of their speed of switching. It may be desirable to use two transition points e.g. switch on at 80°C and switch off at 75°C. A set/ reset latch can be used that is turned on by the in this example the 80°C point and off at 75°C point. This prevents fast switching rates of the fan.

Another alternative is to use a single switch point and feed this to a retriggerable monostable with a time constant that switches the fan on long enough to achieve the desired cooling.

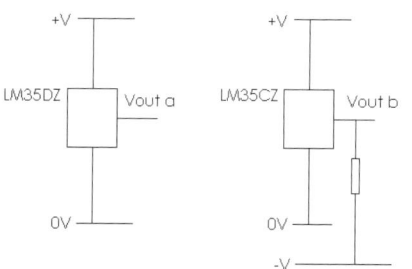

Fig.175 shows the method of using the LM35DZ to measure from 0°C to 100°C or the LM35CZ to measure from 0°C to 110°C. By applying a negative voltage to the output the LM35CZ can measure from -40°C to 110°C. The calculation of the negative biasing resistor is $R\Omega = -V/ 0.00005$. For a $-5V$ negative line this is $5/ 0.00005$. This is $R\Omega = 100000\Omega$. Using the negative sign is not necessary only the relative difference to ground is of consequence in this calculation.

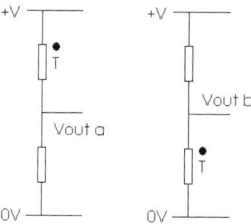

Fig.176 shows the method of using a thermistor for temperature sensing. Most thermistors used for temperature sensing are NTC type. Vout a increases as the temperature increases. Vout b decreases as the temperature increases.

Sound level

The measurement of average sound levels requires some form of amplification and integrator. Transducers with high output such as a crystal microphone may be feasible for direct connection to logic gates.

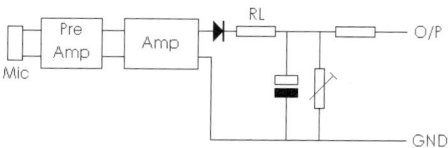

Fig.177 shows the possible block circuit for a sound level input. The microphone feeds a preamplifier and power amplifier. These devices are available as an IC needing few other components. The diode allows the capacitor to charge up but prevents the capacitor discharging through the amplifier. RL is a resistance equivalent to the normal loudspeaker impedance and prevents the amplifier being overloaded because the capacitor is a low impedance device. The preset potentiometer allows the capacitor charge to bleed to ground. The circuit acts as in integrator reacting to the average voltage output over the time period set by the size of the capacitor and the bleed resistor.

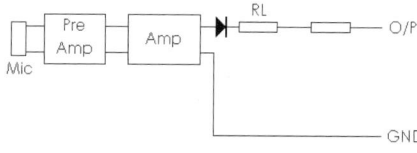

Fig.178 is a similar block circuit but this time the output will follow directly the output of the amplifier. This type of circuit can be used for 'peak level' monitoring or alarm if the output is fed to a comparator.

Inputs 119

QTC 'pills' with rule shown for size comparison.

Light level

Light level sensors usually take two forms. The LDR or light dependent resistor varies its resistance with the amount of light falling upon it. It is used in series with another resistor to produce the same effect as a potentiometer. There are many types of LDR covering a wide range of light conditions. The LDR is a relatively slow device in electronic terms and some can take a number of seconds to adjust fully. They are best used in slowly changing light conditions and the slow response can be exploited because they do not react as quickly to events such as short flashes of light.

The second type is electronic and uses a photo diode or phototransistor, possibly with an integrated amplifier and linear output conditioning. These are very fast devices and can react to light changes often into mega Hz. Depending on the need for absolute or comparative readings, both these outputs can be fed directly to a voltage comparator to give a set level.

Another form of light sensing device is the photovoltaic cell often called a solar cell. This produces a voltage output relative to the light level on the cell. The output can be in the order of 0.45V per cell at 100mA and the cells are joined to give increased voltage or current or both. The outputs can drive comparators directly if a unit is used for measuring. Many of the early photographic meter circuits that did not require a battery were photovoltaic cells driving a meter.

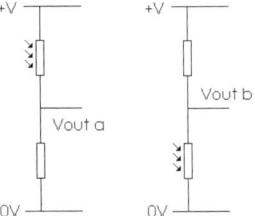

Fig.179 shows a LDR circuit. LDRs decrease in resistance as the light intensity increases. Vout a increases as the light intensity increases. Vout b decreases as the light intensity increases.

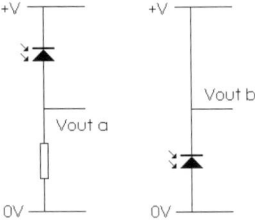

Fig.180 shows a photodiode circuit. The photo diode and phototransistor work in a similar way to the LDR but work on the leakage current and not directly on the resistance. When they are reversed biased very little leakage current flows but as the light increases the leakage current increases.

Pressure

Pressure is usually measured as a result of its effect on other physical movement e.g. the level of liquid in a manometer tube or the displacement of a membrane as in an aneroid barometer. A pressure gauge uses the straightening effect on a curved tube to move a needle by means of links and gears. These all produce an output range. Fixed level devices are also available that operate at a specific pressure an example of this is a steam boiler safety valve. All these units can have sensing devices attached to produce a feedback e.g. the pressure gauge needle could be replaced by an optical disk to give a binary or Gray code output.

The reference to pressure is valid for both positive pressures and for negative pressures i.e. a vacuum.

There have been materials available that produce a variation directly as a result of an applied force. These usually consisted of a composition material that through out the material was a distribution of fine carbon particles. Squeezing the material together caused more of the carbon particles to touch and hence the resistance fell. This is used in weighing scale and similar devices.

Very recently a new material has become widely available this is called QTC - quantum tunnelling composite. This is similar to the carbon impregnated composite. QTC material consists of a flexible compound impregnated with specially treated metal particles that never touch. The operation depends on the quantum mechanic electron wave theory that electrons do not act as a solid particle but as a wave. When these electron waves meet a non conductive interface they do not stop but decay. If the decay is not complete when the wave meets the next conducting area the electron is able to carry on. If the non-conducting areas are large then no electrons will pass and the material is to all intents an insulator. If the material is squeezed so that the non-conducting areas are small then many electrons will pass and the material is to all intents a conductor. The QTC is treated as a variable resistor with a practical range of >10MΩ to <1Ω. Any distortion of the material whether it is squeezing, bending, stretching or twisting will give a resistance variation. In block form it is able to handle voltages to 40V and currents to 10A, temperatures to 120°C and a force rang of 0N to 100N. The material is available as small blocks or as cable.

This looks to be a versatile material where applications may only be limited by the imagination of the user.

QTC can also be used for custom-built switches where the voltage 'on' level is with an OP amp or voltage comparator.

Practical sensors using QTC

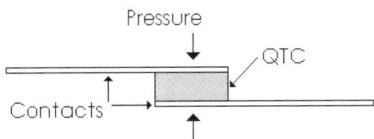

Fig.181 shows a piece of QTC sandwiched between two contacts. When pressure is applied to the QTC the resistance between the contacts falls.

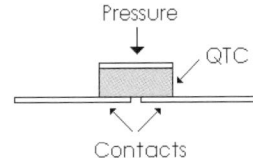

Fig.182 shows a piece of QTC bridging a small gap between a pair of contacts. When pressure is applied to the QTC the resistance between the contacts falls.

Piezo devices

Probably the most common form of this unit is in inexpensive earpieces of the 'crystal' type. The piezo crystal bends when stimulated by a voltage. This phenomenon will produce audible sound when stimulated by an AC voltage in the audio range.

The other common use of piezo crystals was in the pickups of record players for vinyl records. These have now been superseded by the CD, which uses light as the reading means. The pickup used the fact that piezo crystals will also

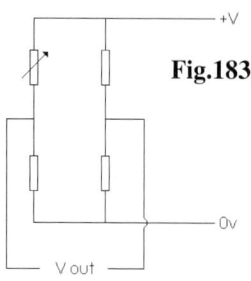

Fig.183

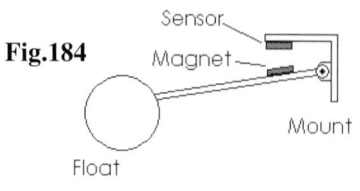

Fig.184

produce a relatively high voltage output when a variable force is applied to cause bending.
A common form of piezo device is the sounder. This is a layer of piezo crystal sandwiched between two brass discs with wires attached. This is normally attached to a thin surface such as the face of a box. The surface will vibrate when an AC voltage is applied to the sounder. These types of devices usually have a resonant frequency in the low to mid audio region.
If this type of device is physically bent or struck a voltage pulse is produced. The output is proportional to the bending effect. The voltage pulse only occurs during physical change. If the device is bent and held a voltage pulse occurs. When the device is released a voltage pulse in the opposite direction occurs.
The output voltage depending on the stimulus is usually high enough to drive voltage comparators directly.
The device can be used to measure vibration at specific frequencies or to act as an impact sensor.

Liquid level

Fixed liquid levels are easily measured by using floats operating a switch of some type. Variable levels can be measured using a float operated variable resistor that can be used in a bridge servo circuit to move a pointer or fed through an ADC to give a digital output.
Fig.183 shows a bridge circuit with a variable resistor as one of the arms. The output can be amplified or could be used to drive a digital or analogue meter directly. The choice of component value will depend on the application and the current required. A good starting point for a 12V system would be 10kΩ for all the resistors.
Fig.184 shows a float with a magnet fitted to the float arm. The sensor can be a reed switch or a Hall effect sensor made waterproof by coating in epoxy resin or similar. A number of sensors could be fitted but the magnets operate over a distance range depending on the strength of the magnet and reed switch or sensor type. Sufficient separation would be required to produce the desired switching action.
Fig.185 shows a magnet inside a float. The float is in a tube that is mounted inside the liquid to be sensed. The sensors shown are moveable to give an upper and a lower limit response.
This same type of set up can be used with optical fibres or light guides if the liquid will

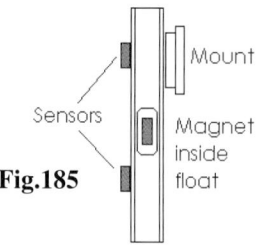

Fig.185

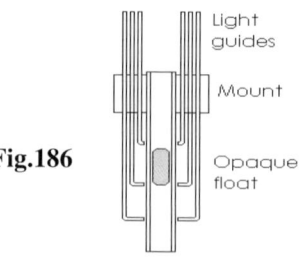

Fig.186

allow the transmission of light. **Fig.186** shows a light guide level sensor system. Three Light guides are shown but in reality a large number can be used. Each receiver light guide goes to an optical receiver. One LED can be used to supply a large number of transmitter light guides. If a large number of light guides are used a number of guides can be blocked at any one time. A simple priority logic system can be used to determine the actual level.

Liquid flow

This is again a secondary result. The usual means is by measuring the rotation of an impeller in the liquid flow. The impeller may have an embedded magnet and the pick up can be a Hall effect or magneto resistive sensor. Using this method the unit does not require any physical connection through the case hence a reduction in the possibility of leakage. If the liquid is clear, optical methods are also possible using light guides or fixed light emitter and light sensor.

If the flow is large a Pitot tube and pressure transducer can be used. The Pitot tube can also be used with a piston or diaphragm to measure the movement and hence the pressure. Another simple solution is to measure the deflection of a vane.

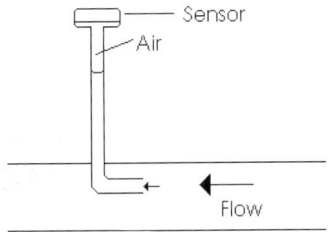

Fig.187 shows a tube where the water flow compresses the air in the tube and the air pressure acts on a sensor.

Fig.188 shows a pivoted sprung vane in the flow the deflection is relative to the flow and hence the pressure of liquid on the vane in the pipe.

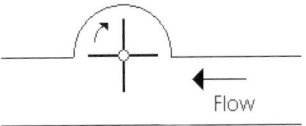

Fig.189 shows a rotating vane sensor. The speed of the impeller depends on the pressure on the vane and hence on the flow.

Air flow

From a measurement point air can be treated as a low-density liquid. All the methods described under liquid flow will work for air. Because of the lower density, larger and differently shaped impellers will be required, as will larger deflection vanes unless the airflow is extremely fast. Pitot tubes and differential pressure sensors are the standard way of measuring airflow and by inference air speed in aircraft.

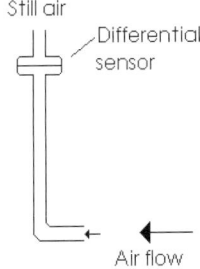

Fig.190 shows a differential pressure sensor measuring the relative difference of still air and moving air and therefore the pressure effect of the moving air.

Flammable gas

The simplest form of gas sensor is the hot platinum wire sensor. Many types of flammable gases and some flammable vapours can be detected. These units consist of a detector and a compensator in a bridge circuit. Combustible gases produce a voltage output relative to the gas concentration. The detector wire in the presence of gas begins to oxidise with a subsequent rise in temperature and change in resistance. This is measured against a compensator wire that has the same characteristics as the detector wire except that oxidation does not occur. A comparator is set to give trip voltage of 23.8mV.

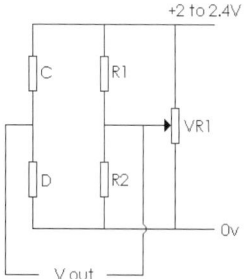

In **Fig.191** D is the detector and C is the compensator. The resistors R1 and R2 that form the other two arms of the bridge should be a minimum of 30Ω and the potentiometer should be 500Ω.

Gear sensors

These are magnetic pickup devices that use the change in inductance to produce an AC output. The device consists of a coil wound around a magnet. They are placed in close proximity to the edge of a gear tooth. Each time a tooth passes the magnet a voltage is induced into the coil. These sensors are easy to make but need a reasonable speed to give an easily readable output. They have advantages of not being in direct contact and they are impervious to dust and dirt unlike most optical sensors.

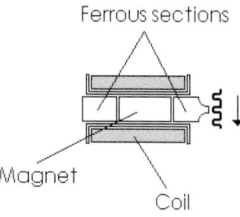

Fig.192 shows a typical inductive coil type gear sensor. They only work efficiently when the speed of rotation is high and a relatively large voltage can be induced.
The Hall effect or magnetoresistive gear sensor has largely superseded them because they are not speed dependent.

Strain gauges

These are used to measure small movement such as stretching or bending. They consist of a copper/ nickel alloy foil pattern on a plastic backing. The plastic backing is often polyester or polyamide and may have a self-adhesive backing. When the gauge is stretched the cross sectional area of foil is reduced and hence the resistance increases. They are often used in bridge circuits and require quite extensive signal conditioning to achieve a usable output.
LVDTs and inductive coil coupling

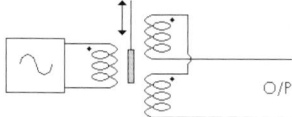

Fig.193 shows a linear variable differential transformers or linear variable displacement transducers consist of three windings around a cylindrical bobbin. The central coil is the primary and the other two coils are the outputs. There is a nickel iron core that can move along the inside of the bobbin. The primary coil is excited by a high frequency signal. The outputs from the two coils are relative in amplitude and phase to the position of the core. At the balance or null point no output occurs.
These devices are often used for measuring small movements such as +/- 1 to 5mm either side of the null point but devices are available that can measure up to +/- 50mm.
LVDTs require quite extensive signal conditioning to achieve a usable output.
Fig.194 shows how it is possible to make a device that works in a similar way to the LVDT by winding two coils around a bobbin. A high frequency AC voltage excites one of the coils

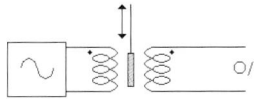

Fig.194

and the output from the other coil is turned into a DC level. When a metal or ferrite core is pushed into the bobbin the magnetic coupling will increase and the output voltage will increase. There is no null point because there will always be some coupling between the primary and the secondary coils. Many variations can be produced depending on the layout of the windings. The device is useful for end position sensing and if well insulated will also work immersed.

CHAPTER THIRTEEN

Light Emitting Diodes

LEDs as indicators

Light emitting diodes are extremely useful as visual indication of an event. The LED is a diode that produces light when a current is passed through it. They are available in a wide range of spectra from ultra violet to infrared. Some are available with multi colour elements and even in built flashing.

Most indicator type of LEDs require a current in the range of 10 to 30 mA. Using ohms law it is a simple matter to calculate the series resistance required. The forward voltage of the LED needs to be taken into account.

The following formula is used

Series resistance =
$$\frac{\text{supply voltage} - \text{forward voltage of LED}}{\text{Current required}}$$

The connection of the LED will depend on whether the circuit is sourcing a current i.e. providing the current required or sinking the current. Sinking current means that the led is connected to the positive rail and the switching device completes the circuit to the negative rail. Sourcing current means that the led is connected to the negative rail and the switching device provides the positive supply.

High power pulsing

Some LEDs particularly infrared types are

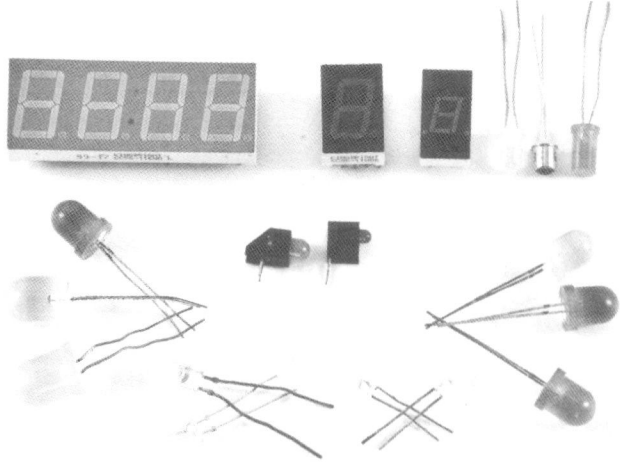

A range of LEDs.

designed to work with many times their normal operating current for very short lengths of time. Some of these LEDS can accept currents in the order of 10A but only for times of ten microseconds. The duty cycle may also need to be taken into account.

Unless definitive data sheets are available from the manufacturer the only safe way is to use a resistor in the LED circuit greater than that expected. The LED is pulsed and an oscilloscope is needed to measure the peak forward voltage across the LED. The resistance is reduced until near the peak forward voltage. An allowance should be built in for any likely power supply fluctuations.

R1 and C1 give the timing of the monostable. The on time is based on 1.1 C R seconds where R is in ohms and C is in Farads. Care must be taken that the input pulse does not exceed the desired output pulse. Multiple LEDs can be driven for even greater output.

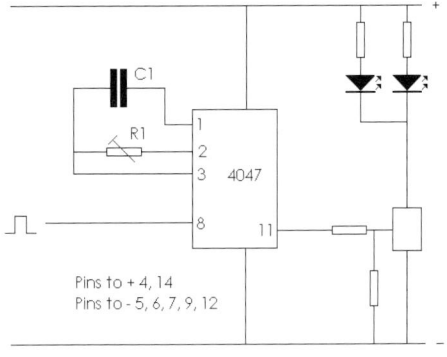

Pins to + 4, 14
Pins to - 5, 6, 7, 9, 12

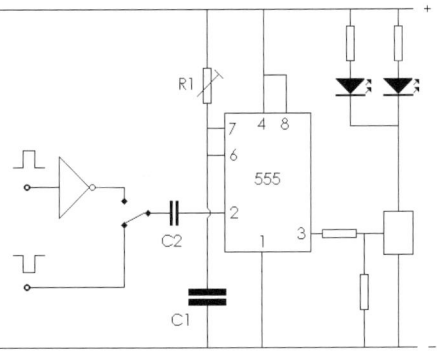

Fig.195 shows a high power LED pulsing circuit.

Fig.196 shows a circuit based on the 4047 positive edge triggered monostable. This diagram is not dependent on the length of the input pulse the triggering occurs on the low to high transition. The 4047 can also be wired as negative edge triggered monostable. The on time is based on 2.4 C R seconds where R is in ohms and C is in Farads. Polarised capacitors cannot be used in this circuit.

Light Emitting Diodes

CHAPTER FOURTEEN

Speed Measurement in the Workshop

Speed measurement in the workshop normally covers three areas, rotational speed, surface speed and transverse speed.

All these areas are connected with the speed of a tool relative to a work piece.

Speed measurement is important in the workshop for safety, tool life and surface finish.

Most speed settings are based on a gear or drive ratio of the speed of the induction motor. This is more difficult with both mechanical and electronic variable speed drives. With a brush motor, the final speed will depend on the loading. It is not possible to use a gear or drive ratio to give any accurate final speed. With these variable speeds, the only alternative is direct measurement of the output. This produces problems particularly at slow speeds.

There are four methods of measuring speed.

Event measurement

This consists of measuring a number of events in a fixed time period. The events may be pulses produced by a magnet and sensor or the breaking of a light beam attached to the rotating shaft. There is one overriding factor, the slower the speed, the greater must be the number of events, or the measurement time must be increased.

This type of measurement suffers from the +/- LSB error. LSB is the least significant bit; this may or may not be counted, hence the +/- error.

Consider a shaft turning at 600 rpm, this equates to a rotational frequency of 10 Hz. To obtain a direct readout of 600, updating at 1 second intervals would require 60 pulses for each revolution. To update at 0.5 second intervals would require 120 pulses per revolution. The error would be 1 in 600, which equates to 0.166% error.

Consider the same scenario with a shaft turning at 60 rpm, this equates to a rotational frequency 1 Hz. To obtain a direct readout of 60, updating at 1 second intervals would require 60 pulses for each revolution. To update at 0.5 second intervals would require 120 pulses per revolution. The error would be 1 in 60, which equates to 1.66% error.

Problems start where it is not possible to have a large number of pulses per revolution. Many revolution counters do not have full displays, they may have a 2 digit display and a multiplication factor of x10 or x100.

Using the same scenario of measuring 60 rpm on a readout with an x10 multiplication factor would require 6 pulses per revolution to obtain a direct readout of 6 with an update of 1 second would have a +/- LSB error of 1 in 6, which equates to 16.6% error. On top of this is the error from the missing display digits. The display may show 600 on a display with an x100 multiplication factor, but the reality is somewhere between 600 and 699.

Gated pulses

This is a variant of the event measurement. The events to be counted are produced by a separate clock. The sensors connected with rotating shaft are used to gate the events to a counter.

This type of measurement suffers from the +/- LSB error, but not to the same level as the direct event measurement because of the larger number of pulses in the time period.

The principle is that a clock that can run at high speed is gated to a counter by sensors actuated by the rotating shaft. As few as one sensor can be used and can sequentially actuate a start gating, stop gating, transfer result etc. The problem with this system is the output that is inversely proportional to the rotational speed, i.e. the slower the shaft is rotating the more clock pulses are allowed through. If clock speeds are high, the +/- LSB error will be insignificant. The main problem is the display, which will require a processor or at the least a lookup table held in memory or on an EPROM.

The gated pulse count is very good for go/no go testing of speeds, such as induction motors that could be overloaded or for starting and up to speed sequences. The normal method is to feed the clock to a charge pump device. Charge pump devices are generally more efficient at high clock speeds. These pulses are converted to a DC voltage and used with a window comparator to make sure the motor is running at a speed within a set of parameters.

Time interval

This is based on the direct correlation of speed and the time interval between two events. This is a direct reading, which has no inherent errors. This is probably the most accurate and simplest method of measuring speed. The accuracy is a function of the measuring clock. A sensor is used to initiate the start of a clock count, the same or another sensor is used to stop the clock count. The measurement is the time interval for one revolution or part of a revolution. The most accurate is to use the same sensor to measure the start and stop for a full revolution. This method prevents errors due to sensor misalignment or angle measurements. With high-speed crystal controlled clocks the accuracy in practical applications can be in the order of milliseconds or even microseconds. Some form of processor will be needed to control the operation and convert the result to a usable display. As an example a count of 100 milliseconds to complete a revolution equates to 10 Hz and a speed of 600 rpm. This system of measurement does not have problems with speeds as low as 1Hz, which is 60 rpm. The accuracy remains a factor of the clock accuracy.

Indirect measurement

This is where a comparison of speed is made against a known speed or frequency.

Rotational speed

This is the speed at which an object is rotating. It has no mathematical connection with the size of the object.

This is probably the easiest speed to measure. Sensors and actuators provide an output that can be directly manipulated to provide an output in rpm. Both optical and magnetic sensors can provide good outputs but many forms of optical sensor are sensitive to dust and dirt.

Calibration discs

Rotational speed can also be measured or calibrated by using a disc with a number of light and dark segments. This can be used with a stroboscope or with some known frequency light source. A fluorescent light will produce a 100Hz light flicker at 50Hz and at 120Hz on a 60Hz Mains AC supply. When the disc appears stationary the disc is at the speed of the flashing light source – or unfortunately a multiple of the light source. Consider a two-segment disc that appears stationary at 6000RPM at 50Hz. This disc would still appear stationary at 12000RPM although in effect you are only seeing every second rotation. This may not be a problem with synchronous AC motors where by means of

Calibration disc segments alternate black and white segments		
Segments	RPM at 50Hz	RPM at 60Hz
2	6000	7200
4	3000	3600
6	2000	2400
8	1500	1800
12	1000	1200
16	750	900
24	500	600
48	250	300
120	100	120

calculation of the motor speed range and gearing it is possible to be within the speed range of the calibration disc. With DC motors an educated guess may be the starting point if no other measuring means are available. Despite the negative point raised, calibration discs are still useful.

Calibration discs can easily be produced on a PC. They consist of segments alternately coloured black and white. The table gives useful range of speeds.

Calibration discs are useful for setting up a system but if continuous readings are required then some other means needs to be used.

Surface speed

There is a direct connection between the diameter of an object, the rotational speed and the surface speed of an object. The larger the diameter of an object for a fixed rotational speed, the higher will be the surface speed.

The normal way of measuring surface speed is with a contact disc of known circumference. Because the circumference is known, the distance the wheel travels is directly related to its rotational speed. A small processor is used to calculate the surface speed in terms of the desired output e.g. feet per second, metres per minute etc.

The major problem for the home builder of this piece of equipment is the accuracy needed in the circumference of the measuring disc. These are available commercially but are intended for use with specific equipment and are expensive. With the use of processors, it is possible to use multiplication factors for measuring discs, and provided the diameter can be measured accurately, the circumference can be calculated and a correction factor applied.

Transverse speed

This is the crosswise movement relative to a fixed point. It is normally used in conjunction with a rotational speed. An example of this is the automatic feed of a cutting tool in a lathe turning operation. There is a relation between the crosswise movement of the tool and the rotational speed of the work piece.

This measurement only becomes significant on a machine that has independent drive for traverse or cross feed. On the lathe these drives are usually related to the headstock speed. On a milling machine or grinder these are normally independent drives. With stepper motors the speed is a factor of the step speed, with DC motors this is a function of the rpm of the motor. It would be a simple procedure to calculate the gear ratio and then use a lookup table, either manual or processor controlled, based on rotational speed of the motor. If a variable ratio gearbox is fitted, then it is probably simpler to use direct reading using similar techniques to surface speed measurement.

CHAPTER FIFTEEN

Power Supplies & Regulators

Care must be taken with any unit connected to a mains supply. All mains connections must be insulated and any enclosures must be designed to prevent fingers accessing any danger area. When used in a workshop there are other dangers such as swarf and cutting fluids that could short out a power supply.

Design of power supplies is a book in its own right but simple unregulated power supplies can be easily made and by using standard voltage regulators regulated power supplies are easily built. These are available in a range of voltages and current and are usually adequate for powering most small to medium electronic projects.

The high current drivers are powered directly from the unregulated power supply.

Relative voltages

Although reference is made to plus and minus voltages this is purely a notional reference. The notion of plus or minus is a term relative to a point of measurement. The nearest that is likely to be a reasonably fixed reference is earth, which as the name implies is a reference to the surface of the earth. This is sometimes used synonymously with the term ground. Ground in the context of this book is the reference from which other voltages will be measured. Because of safety in units that are not earth isolated the

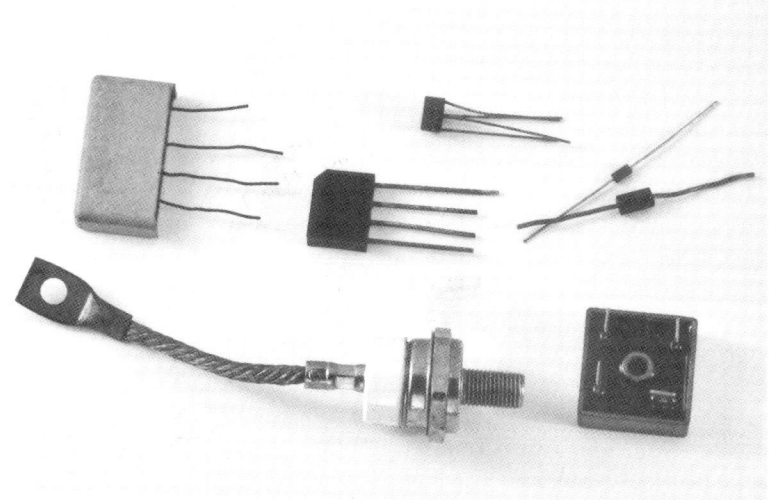

A range of rectifier diodes and bridges. The current rating of the devices shown are from 1-300 amps.

chassis of a unit will usually be used as ground and this ground will be tied to earth via the mains lead hence the synonymy.

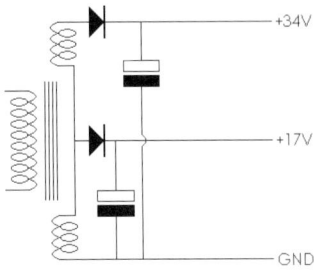

Fig.197 shows a basic power supply based on a 12 – 0 – 12 transformer. This gives outputs of +17V and +34V relative to GND.

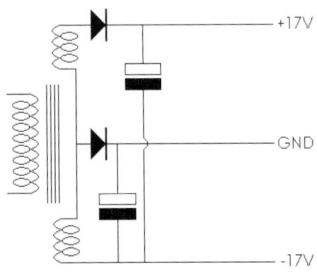

Fig.198 shows exactly the same power supply but the previous +17V output is tied to ground reference. This results in +17V and –17V outputs. The circuit is fundamentally unchanged only the point of reference is moved. Any components attached to the circuit see exactly the same voltages as previously because they all work relative to their own notional ground. In many circuits, particularly those for battery use, the notion of plus and minus is used instead of plus and ground because of the battery voltages relative to each other. Whichever way you choose will make no difference to the working of the circuit except that in a circuit that is earthed a reference to this will be needed. The choice of this may be affected by factors such as interference suppression.

Unregulated supplies

All the power driver projects in this book require only a basic power supply consisting of a transformer, bridge rectifier, capacitor and necessary safety devices. Regulated supplies will be needed for logic circuits but these only usually require low power.

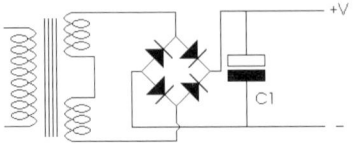

Fig.199 shows a basic power supply. The transformer and the regulator must be chosen to be suitable for the voltages and currents needed. C1 is the smoothing capacitor and must have a voltage rating greater than the peak input voltage. The DC voltage will approach the peak of the input AC voltage. This is $\sqrt{2}$ multiplied by the RMS input voltage. This approximates to the following voltages.

6V RMS = 8.5V DC.
9V RMS = 12.75V DC.
12V RMS = 17V DC.
18V RMS = 25.5V DC.
24V RMS = 34V DC.

By how much the rating should exceed the expected voltages and currents is always a matter for argument. Manufacturers often only overrate by a few percent. This is because higher rated components often cost a little more. For the person making a unit for themselves a 50% rating will usually only add a few pennies to the cost but increase the reliability. Military electronic units are often overrated by as much as 400% to ensure reliability.

The only component that may not need to be overrated by a large amount is the transformer but this will depend entirely on the circumstance that the unit is used under. If the unit is subject only to very intermittent use then the situation is totally different to the unit that is in use all day and may be subject to being occasionally overloaded. Most manufacturers rating are the maximum not the norm.

Transformers are available with single or multiple windings. The most common type will be denoted as voltage – 0 — voltage e.g. 12 – 0 – 12. This means there are two windings of 12V. The two windings may be configured as a centre tap or two independent windings. If the transformer is described as 2 x 12V, this indicates two independent windings.

The transformer may be described in terms of current available from each individual winding or in terms of VA. VA is voltage multiplied by current e.g. output voltage 2 x 12V, output power 120VA, and output current 2 x 5A. This indicates two windings supplying 12V RMS at 5A each i.e. 10A in total at 12V this is 12 x 10 = 120VA.

The advantage of transformers with a pair of equally rated windings is that they can be wired in series to produce double the voltage or in parallel to produce double the current of the individual windings e.g. output voltage 2 x 12V, output power 120VA, and output current 2 x 5A. This can be wired to produce 24V at 5A or 12V at 10A.

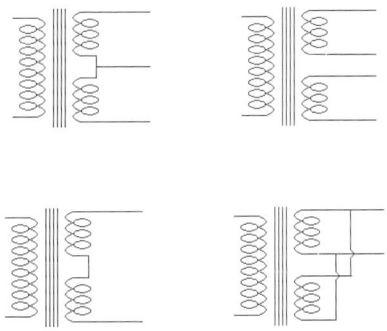

Fig.200 and **Fig.201** show various transformer configurations.

Instead of a bridge rectifier, two separate rectifier diodes can be used for a power supply. The DC voltage will be equivalent to √2 multiplied by the voltage of one of the windings.

By using positive and negative pairs of diodes both negative and positive relative to the centre tap DC voltages can be produced (this diagram is in effect a bridge rectifier with the diodes shown individually). An encapsulated bridge rectifier could replace the four diodes. Single diodes could be used instead of paired diodes if the greater ripple is not a problem. A pair of diodes to the same output line will produce an output ripple at twice the mains frequency because both the positive and negative half cycles are rectified. A single diode connected to the output line will produce an output ripple at the mains frequency because either the positive or the negative half cycles are rectified depending on

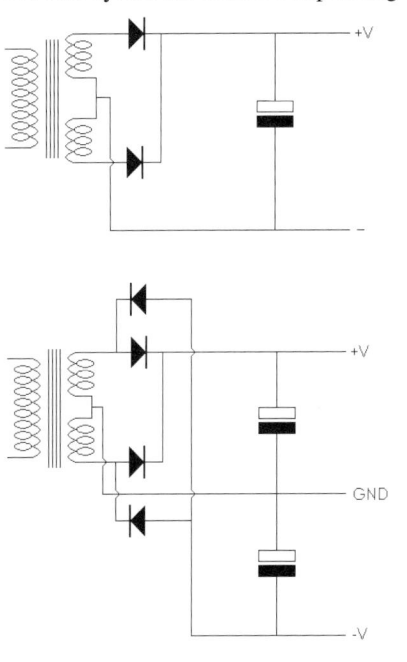

the direction of the diode. **Fig.202** and **Fig.203** show various transformer and rectifier configurations.

Fixed voltage regulators

These are normally available in the following positive voltage outputs +5, +6, +8, +12, +15, +18 and +24 with negative voltages of –5, –12, and –15.

The current range is usually 0.1, 1 and 2A with a limited range of voltages at 5A.

There are also a small number of variable voltage devices but a standard device can be set to a higher voltage by putting a voltage offset on the ground reference pin.

There are a few regulators referred to as low dropout. Most regulators require a supply typically 1.5 to 3V above the regulated voltage.

Low dropout regulators only need a supply typically 0.5 to 1V above the desired regulated voltage. This makes them particularly useful with battery supplies.

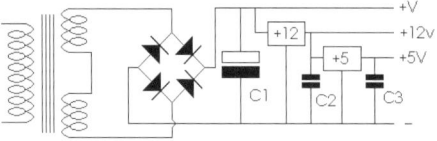

Fig.204 shows a typical power supply where regulated voltages are tapped off the main power supply. Regulators will produce heat relative to the current being taken and to the voltage drop from the main DC voltage. Dissipation can be shared over a number of regulators by reducing the voltage in steps. In this example the current needed by both the +12V supply and by the +5V supply will pass through the +12V regulator. Capacitors C2 and C3 provide 'smoothing' for the regulator outputs.

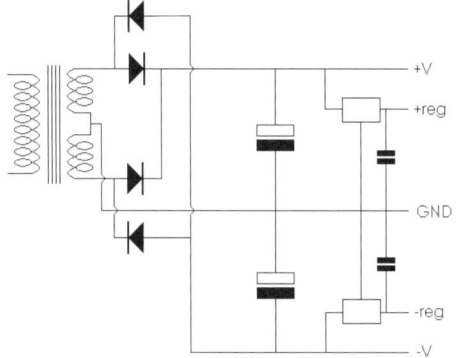

Fig.205 shows the same circuit as previously but set up to provide regulated + and − DC supplies from the main DC voltages. This type of circuit is often used for OP amps requiring dual supply lines.

Specialised power supplies

Fig.206 shows a power supply for a dual voltage stepper motor drive using 'N' channel MOSFETS. The circuit looks complex but is simply a set of small blocks joined together.

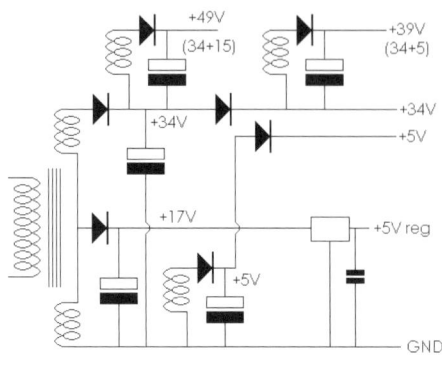

Fig.206

The basis of this is a 12 – 0 –12 toroidal transformer. This provides the ground reference, the +17v ($\sqrt{2}$ x 12) and the +34V ($\sqrt{2}$ x 24). The +5V supply is from an over winding added to the transformer. The +49V is from another over winding to give 15V but to save on wire turns it is made ground reference to the +34V hence the output is +49V. This is used for the electronic voltage control for the voltage level switching. The +39V is from another over winding to give 5V but to save on wire turns it is made ground reference to the +34V hence the output is +39V. This is used for the high side driver gate voltage. The +17V output is used to give +5v logic via a regulator and to provide the low side driver gate voltage.

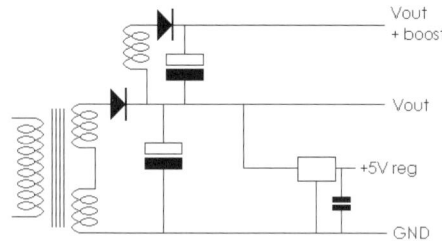

Fig.207 shows a similar power supply used with 'H' bridge and PWM drive for DC motors. The output can be set to that required by the motor and the boost can be set to that required by the high side drivers. The low side drivers can be driven from Vout or a suitable tap or over winding as previous.

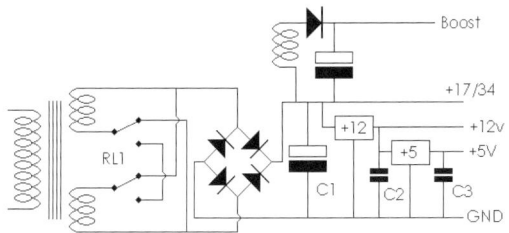

Fig.208 shows a dual voltage power supply with all the voltages needed for a PWM 'H' bridge drive and logic. Because the boost voltage is ground referenced to the +17/34 output voltage it will automatically be Vout plus boost. A relay is shown as the method of changing voltage because this formed part of an automatic sense system when a particular unit was plugged in. For non-automatic systems this relay could be replaced with a suitable switch.

Increasing the voltage O/P of fixed regulator voltages

Because most of the simple fixed regulators are referenced to ground, raising the ground reference can be used as a means of increasing the voltage output from the regulator. This can be by using the forward voltage drop of a diode or a number of diodes in series or from the voltage across a zener diode.

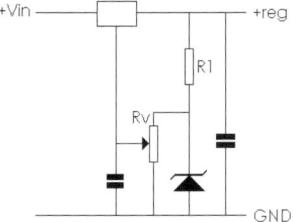

Fig.209 shows a variable voltage ground reference using the voltage across a zener diode. R1 prevents an excess current flow. The voltage is tapped off across a potentiometer. This means that the ground voltage offset can be from zero to the zener voltage. The offset is limited in this type of circuit because the zener voltage needs to be less than the regulator output. The output tends to be slightly less regulated because of the two voltage references but is adequate for most logic applications.

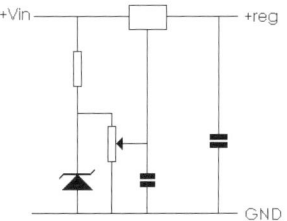

Fig.210 provides a greater output range that is mainly dependent on the voltage in. This diagram provides a wide range of voltages but because the zener reference is from an unregulated source there is more tendencies for the voltage to float slightly but again is usually adequate for most logic applications.

The above regulator circuits are all classed as linear control. They work by changing the conduction of the pass element. Although the regulation is generally good and there is little output ripple hence the need only for small capacitors the down side is the efficiency and heat dissipation.

PWM power control

In a similar way to motor control the PWM circuit can be used to supply a variable voltage. If the motor is replaced by a capacitor and no load is added the capacitor will charge up to the peak voltage of the input. If a load is added the voltage across the capacitor will be dependent on the current taken and the PWM ratio. Voltage measuring and feedback circuits can be added to produce a stable voltage by adjusting the PWM ratio.

Switch mode voltage control

The switch mode power supply came about because of the need for inexpensive, lightweight, efficient and adaptable power sources. Most low voltage linear power supplies use a drop down isolating transformer to provide the necessary voltages. The transformer is usually the most

A range of voltage regulators.

expensive and heaviest part of any power supply. The switch mode power supply works in a similar way to PWM in that the voltage output is dependent of the switch on time ratio to switch off time of the power control section. Using digital switching means that the dissipation is low in the output components because the output driver is either fully on or fully off. These states give a theoretical zero dissipation, but in reality the output devices have a transition time between states and a small resistance that causes heat to be produced.

Many commercial switch mode power supplies work by rectifying the AC mains to a DC voltage and then switching this DC voltage on and off to give the required output voltages. Voltage measuring and feedback circuits are added to produce a stable voltage by adjusting the switching ratio.

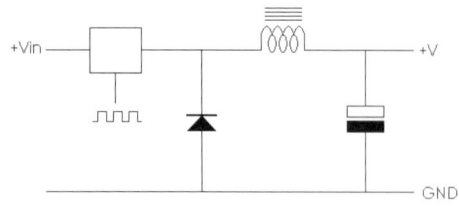

Fig.211 shows the voltage switching components of the switch mode circuit.

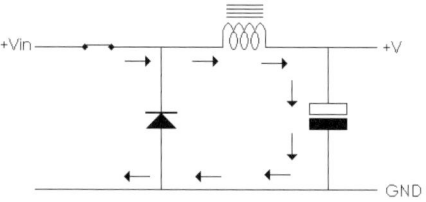

Fig.212 shows the effect when the switch is on. The inductor or choke acts in the same way as the coil of a motor and 'resists' the rise in voltage. The inductor is usually large enough so that it does not become fully saturated at the frequencies used for switching. The capacitor charges by the path shown.

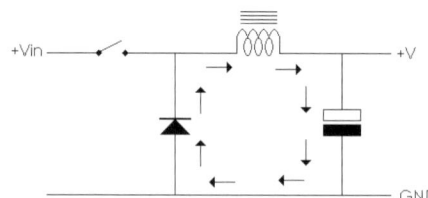

Fig.213 shows the effect when the switch is off. The energy stored in the inductor discharge via the diode and the path shown increasing the voltage on the capacitor.

Because of this switching it is possible to obtain numerous voltage outputs from one input voltage. Inverted output voltages can be produced, as can voltages that are higher than the input. The greatest advantage of the switching power supply is the ability to produce regulated low voltage power supplies capable of providing very high currents.

The circuit description shown is for a switch down regulator but there are two other configurations for this type of regulator. All three types use the stored energy in the inductor to produce the output. Block diagrams of the three types are shown for comparison.

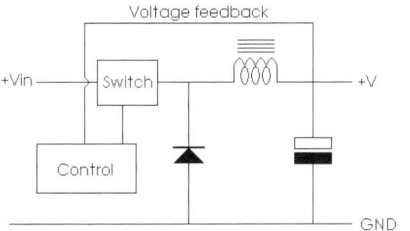

Fig.214 shows the configuration for a switch down regulator

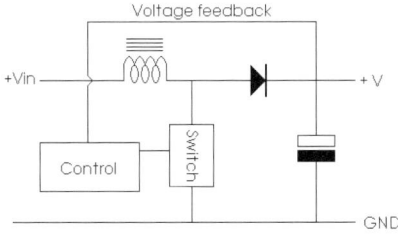

Fig.215 shows the configuration for a switch up regulator

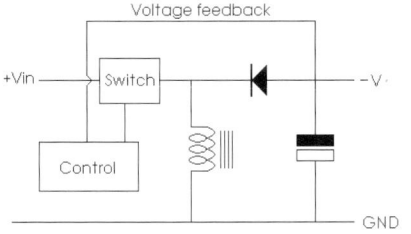

Fig.216 shows the configuration for an inverting regulator

Both step down and step up voltage regulator ICs are available with data sheets available showing standard set ups

I do not consider it advisable for the amateur producer of power supplies to use rectified mains input. It is far safer although a little more costly to use a step down transformer.

The TL497

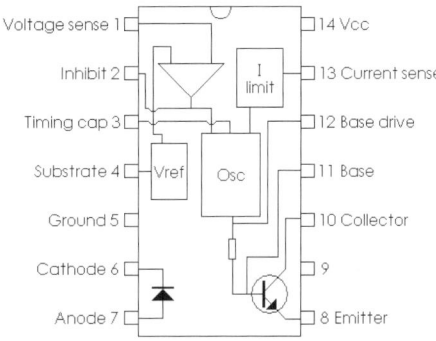

Fig.217 shows the pin out for the TL497 switched mode power supply IC. The current rating of the inductor sets the current that can be taken. The IC can supply up to 0.5A using the internal switch and diode. With an external switch and diode the current can be up to 5A. The input voltage is 7V to 15V and the maximum output voltage is 35V. The 1Ω resistor forms part of the current limiting and short circuit protection. The current limit occurs when the voltage drop across the resistor and therefore the difference between pin 14 and pin 13 is about 0.7V. Therefore with a 1Ω resistor the current limit is about 700mA. The resistor can be changed to provide an appropriate current for the components being used. If an external switch is being used the drive can be taken from the transistor base drive on pin 12 or the emitter of the internal transistor on pin 8 can be grounded and the collector output on pin 10 can be used for the drive. The voltage setting is from the reference voltage and the ratio of R1 and R2. The reference voltage is shown as 1.2V but can from

Power Supplies & Regulators 137

about 1.1V to 1.3V. If the output is critical a variable resistor or a potentiometer can be used to form part of the circuit of R1 and R2.

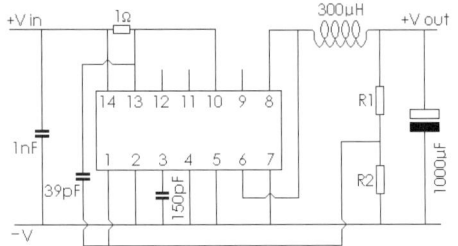

Fig.218 shows the circuit for a TL497 step down regulator.
The output voltage is 1.2(1+R1/R2).

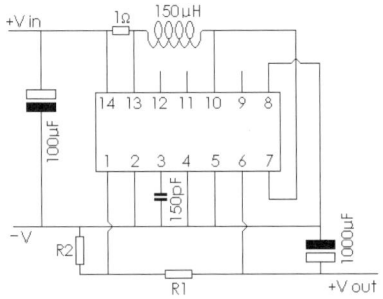

Fig.219 shows the circuit for a TL497 step up regulator.
The output voltage is 1.2(1+R1/R2)

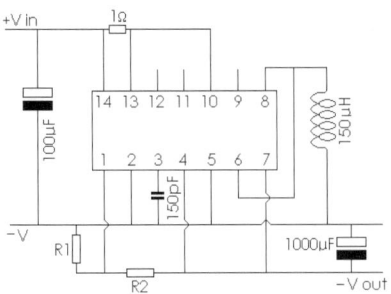

Fig.220 shows the circuit for a TL497 inverting regulator.
The output voltage is -1.2(1+R1/R2).

The LM2577

Fig.221 shows the layout of one version of the

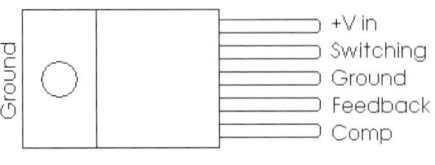

Fig.221

LM2577 step up regulator. It is available in a number of other packages but the type shown is convenient for mounting to a heatsink. The tab is not isolated from the internal circuitry.

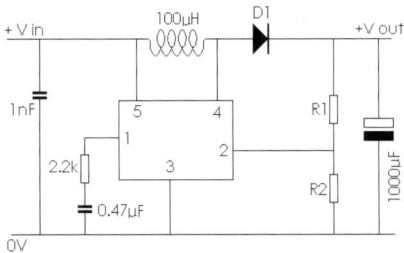

Fig.222 shows a circuit for the LM2577 step up regulator. It is described as a step up regulator but step down and inverted outputs can be obtained by using a ferrite transformer instead of the choke. It is available in three forms, a fixed +12V output, a fixed +15V output and an adjustable output. The IC can accept inputs of 3.5V to 40V and provide output currents up to 3A. The diode D1 needs to be a Schottky type diode e.g. a 1N5821 for output voltage up to 50V and a fast recovery diode for voltages up to 100V.
For the adjustable output the V out is 1.23(1+R1/R2).

The ICL7660

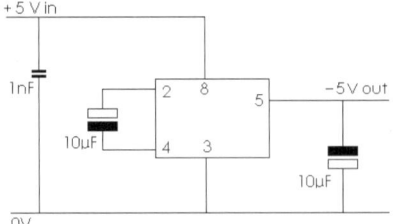

Fig.223 shows the ICL7660 switched capacitor voltage converter. This IC is capable of being

Electromechanical Building Blocks

used as a voltage doubler in 1.5V and 3V battery circuits. The current output is very low, typically in the region of 110mA. The most useful application is for the provision of –V supply from a +ve supply for OP amps in battery circuits.

Stepped voltages

Stepped voltages are useful when driving stepper motors and solenoids. The voltages chosen and the current required will determine the layout. Higher voltage can often be taken directly from the unregulated power supply. Multiple supplies can be taken from the same source.

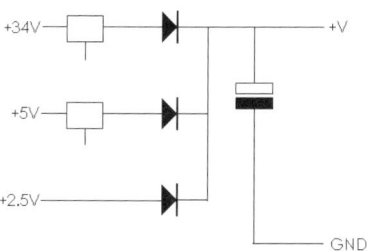

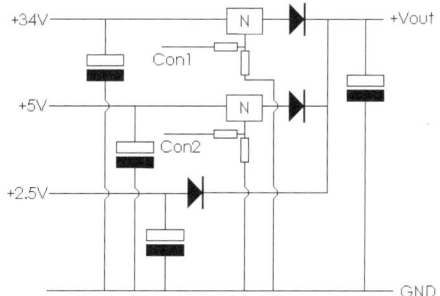

Fig.225 shows a practical version of the stepped voltage drive for three voltage levels. The 2.5V level is always present to the output. If this was required to go to zero another N' type MOSFET driver would be required and configured in a similar manner to Con1 and Con2. The inputs Con1 and Con2 are boosted drives to the 'N' type MOSFET drivers. Depending on which of these are switched on will determine the output level. The diodes isolate the power inputs from the power output. Appropriate levels of drive are needed as discussed previously because the two 'N' type MOSFET drivers are in effect high side drivers. Boost voltages would need to be in the order of fifteen volts and zener protection on the driver gates would be appropriate.

Fig.224 shows a three level circuit for a stepper motor. The transformer has windings of 1.75V, 3.5V and 24V. These equate to smoothed and rectified voltages of 2.5V, 5V and 34V. The voltages are used for hold voltage, normal phase voltage and start of phase step voltage respectively. Drivers are used to supply current through individual diodes to a small capacitor. The diodes will drop a voltage equivalent to their forward voltage but this is negligible compared to the voltage levels. The capacitor is chosen to be small enough that the voltage will rise and fall quickly when a load is taking current. Only two control drivers are required if the lowest voltage will be the hold voltage. If a zero voltage control is required it is necessary to fit a control driver to the 2.5V line to make it possible to switch this off. The highest voltage switched on will appear on the capacitor irrespective of the other voltages being switched on or off.

Extra voltages from transformers

Transformers come with fixed voltage settings. It is rare to find a transformer that will supply all the required voltages for a design unless the current required are relatively low and regulators can be used. Multiple transformers can be used but often add to cost and complexity.

It is relatively easy to add extra windings to some transformers. Some transformers are supplied as a core with a pre wound mains winding and an empty bobbin for your own windings. The VA range tends to be limited and they can sometimes be difficult to source. Most large standard ferrous cored transformers have sufficient room for small extra windings over the existing bobbin. The output voltage will be a matter of experiment but four turns per Volt is a good starting point. Toroidal transformers can accept more windings around the core. The windings should be spread out as evenly as possible to distribute any heat produced. Any windings added to a transformer add to the total VA of the transformer. Extra secondary windings will impose a corresponding load on the primary winding. With the stepper motor and solenoid driver circuits the voltages taken are mainly on an either/ or basis and therefore do not increase the loading. These factors must always be taken into account when choosing a suitable transformer. The toroidal transformer core is of a higher magnetic rating than ferrous cores. This usually means that less amp turns are needed to saturate the core. The output voltage will be a matter of experiment but two and a half turns per volt is a good starting point.

With any winding added to a transformer it is important that none of the turns are shorted. All windings should be well insulated. A shorted turn will attempt to pass enormous currents usually causing the wire to melt. With toroidal transformers in a metal box the mounting bolt will create a shorted turn if the bolt is fastened to the base and comes in contact with the top of the case.

CHAPTER SIXTEEN

Power Supplied From Batteries

With units intended to be portable it may be that the only power source will be a high capacity battery. This may be in the form of a car battery or NiCd cells. In either case the voltage is likely to be a maximum of 24V. This may not be a problem for the motors being used but controllers and high side drivers may require supplementary voltages.

These voltages can be obtained by adding supplementary batteries of a capacity to match the application. This approach does work but there is always a problem with charging and with being sure that the supplementary batteries do not run out before the main battery. This can lead to loss of control of a unit and possible catastrophic consequences.

It is safer and more convenient if the supplementary voltages can be obtained from the main battery especially if the main battery has a monitoring system for the voltage level.

Boosted battery supplies

Boosted supplies imply that the voltage is higher than the original battery supply but it may also be necessary to supply a negative supply for an IC.

Boosted supplies are usually low current. It may only be necessary to supply a few milliamps to drive electronic devices.

There are a number of approaches to supplying supplementary voltages with enough current capacity for driving for example high side MOSFET output stages or supplying negative supplies for OP amps or processors.

The simplest is the ubiquitous 555. This can be used in voltage multiplying circuits to obtain two, three or many times the input voltage. The basis of multiplying circuits is a chain of diodes and capacitors that add the original input voltage to the previous voltage. The output voltage depends on the length of the chain. Positive voltages relative the +ve rail and negative voltages relative to the 0V rail can be obtained. Voltages can also be obtained with the 555 by using a small transformer. The design and matching of the transformer is more difficult than using the diode and capacitor network.

There are also many dedicated integrated circuits that can supply both boosted positive and negative supplies. The most common particularly in the amateur market is the TL497. The data sheet for this was originally published in June 1976 and revised in February 2005. The longevity of the device is probably down to the ease of use, reliability, low cost and is obtainable easily. The TL497 and a number of other circuits equally suitable for mains or battery use are described in the power supply section.

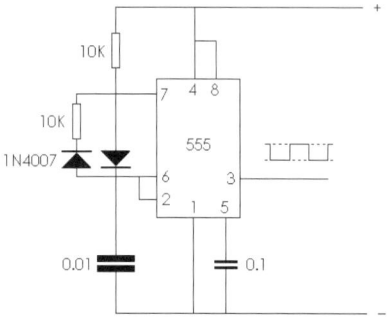

Fig.226 shows a practical driver for boost voltages using a low power 555. The output is a square wave. With the component values shown the output is about 720Hz. It is possible to use other types of square wave or AC drivers but the circuit shown is inexpensive and under test with the following circuits produced an output current in the order of 50mA, albeit with a voltage drop in the output of about 25%. This current is well in excess of that needed for applications that it is envisaged the circuit would be used in. The usual way that the voltage boost is used is to feed the output to a 'storage' capacitor and then to a regulator. A voltage would be used that is in excess of the required regulated output.
This circuit is also capable of driving a small transformer but this is more complicated because of the need to match the output frequency with the impedance of the transformer.

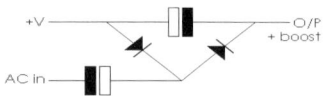

Fig.227 shows a simple positive voltage multiplier. It is classed as a voltage doubler but losses and inefficiencies produce an output voltage that is less than double the input. With the 555 driver circuit and a 12V positive supply

the output was measured at 22.7V. This is a boost of 10.7V and in percentage terms is a boost of 89%. This voltage loss equates closely to the voltage drop of the diodes. The diodes used were all 1N4007. These are efficient at the low frequency being used and have a PIV of 1000V. The voltage rating of the capacitors used should be selected to suit the voltages present. Capacitor polarity is important and should be noted from the circuits. This polarity varies with stages and with application.

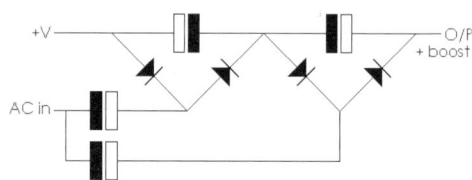

Fig.228 shows a progression from the simple positive voltage multiplier. It is classed as a voltage tripler but losses and inefficiencies produce an output voltage that is less than three times the input. With the 555 driver circuit and a 12V positive supply the output was measured at 33.4V. This is a boost of 21.4V and in percentage terms is a boost of 178%. This percentage of 178% is twice that of the 89% of the doubler circuit. It is feasible to add extra stages if higher voltages are required. Each stage would theoretically with this example add a boost equivalent to 89% of the voltage.

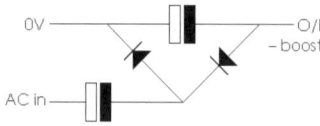

Fig.229 shows a simple negative voltage boost circuit. With the 555 driver circuit and a 12V positive supply the output was measured at − 10.7V. This circuit equates to the positive boost

circuits in all respects except the output polarity.

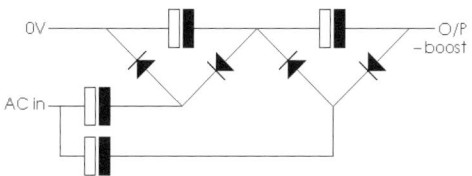

Fig.230 shows a simple negative voltage doubler circuit. With the 555 driver circuit and a 12V positive supply the output was measured at − 22.4V. This circuit equates to the positive boost circuits in all respects except the output polarity. The circuit is called doubler but looks like the positive voltage tripler because measurement for positive boost voltage are with respect to the positive rail and in negative boost circuits are with respect to the 0V rail.

Battery voltage monitors and cut outs

Most of the descriptions relative to battery circuits apply equally to most power supplies. The descriptions are in this section because battery equipment is often more susceptible to low voltage and high currents than mains powered equipment.

Battery voltage monitors and cut outs extend from simple metering to controlled shut down procedures in the event of low battery voltage. For the purposes of the following circuits it is necessary to differentiate between voltage measuring and voltage monitoring. Voltage measuring is the reading of an actual voltage that may change with time or with battery use. Voltage monitoring is the use of that voltage reading to trigger an external event. The event may be the lighting of an LED or it could be a warning or a shut down sequence.

The simplest form of voltage measurement is to use a moving coil voltage meter. The disadvantage with most meters of this type is the linear scale starting at zero. With a car type battery the portion of the scale that is of interest is between about 10V and 15V. This equates to about one third of the full scale and means that readings are more difficult because of the bunching.

Digital voltmeters are a better solution. These are available as panel mounting units. With the increasing use of the digital multimeter it is now possible to buy a low-end version for below £5. This contains everything needed to produce a reasonably accurate panel meter with a small amount of work.

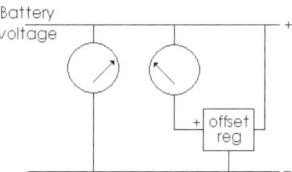

Fig.231 shows the effect of using a regulated offset voltage to produce a shifted display meter. The left hand meter is connected directly across the voltage.

A meter shows a relative not an actual voltage i.e. the voltage shown is the difference between the points of measurement. If a positive voltage is placed at one side of a meter then the voltage shown is the difference between this voltage and the positive supply voltage on the other side of the meter. The offset voltage can be produced by a regulator or by a zener diode and resistor. A meter must be chosen that matches the range required. A 5V full-scale deflection meter with a

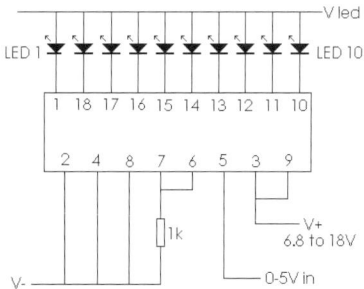

10V offset will display a voltage range from 10V to 15V.

This type of display is not suitable for a very

narrow scale near the full battery voltage. When the battery voltage falls to or near that of the offset voltage the indication on the meter will be erroneous because the offset voltage will also fall.

Simple bar or dot displays can be made using LEDs. These have the added advantages of being robust and different colour LEDs can be used for separate sections of the display.

There are a number of dedicated display drivers such as the LM3914 and LM3915 available. The LM3914 is a 10 digit LED driver with a linear response that can give a bar or dot display. The LM3915 gives a logarithmic response.

The LM3914 is a very versatile device. Fig.232 shows the simplest possible configuration for a bar display with a 0V to 5V input i.e. 10 LEDs at 0.5V per step equals 5V. This range is adjustable but it is usually easier to change the input range via a potentiometer to be a 0V to 5V input. If the required voltage to be measured is 12v, then a potentiometer is set such that an input of 12V produces an output of 5V. This means that each LED step equates to 1.2V based against the original 12V. The resistor on pin 7 adjusts the brightness of the LEDs. It is a wise precaution to connect a 2.2µF capacitor between the positive voltage for the LEDs and the ground at pin 2. The resistor ladder network for the switch reference points is an internal function of the device. This means that only steps of one tenth of the range can be used.

The manufacturers data sheet is excellent for this device and gives a large number of illustrations of different and extended configurations.

for a dot display with a 0V to 5V input. Pin 9 controls the bar/ dot function. A switch can be used to connect pin 9 to the positive rail or to pin 3, to make the function selectable.

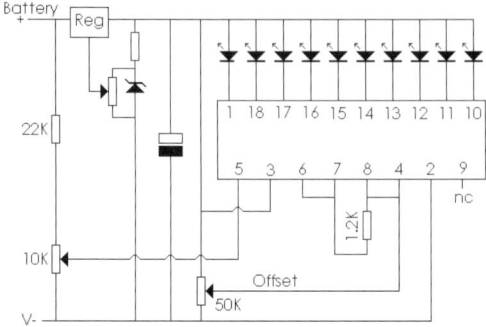

Fig.234 shows the LM3914 as a car battery monitor. The regulator should be a low dropout device and is set up to give a voltage that is two volts less than the lowest voltage to be monitored. This is described in the section on power supplies. A practical measuring range is 10V to 14V. The internal regulator is 1.25V to measure 5V therefore an offset of 10V is approximately 2.5V. This is not exact because of the other resistances in the circuit. The unit is easier to set up if a variable power supply is available. To calibrate practically set the offset to the appropriate calculated value then set the voltage in to 14V and turn the potentiometer connected to pin five until the highest LED just turns on. Check the range by adjusting the input voltage. Adjusting the offset can make alterations to the voltage range. Because the offset and the input voltage are interconnected, the input voltage will need to be adjusted.

Fig.235 shows a bar graph output made using a series of voltage comparators. Only three OP amps are shown for clarity. Adding more comparators will extend the number of outputs. This is more complicated than using a dedicated driver such as the LM3914 but individual trigger points can be set up for each segment of the display by adjusting the values of the ladder resistors. The +ve rail should be fed from a

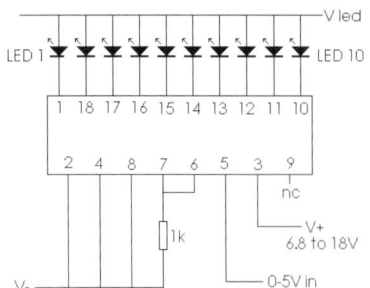

Fig.233 shows the simplest possible configuration

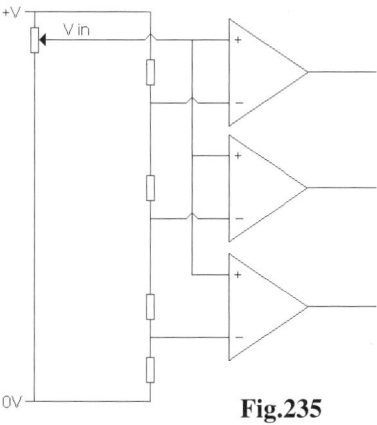

Fig.235

output of the higher stages must be high enough to fully switch off the low stages allowing for the voltage drop at each diode i.e. higher than the reference voltage at that stage.

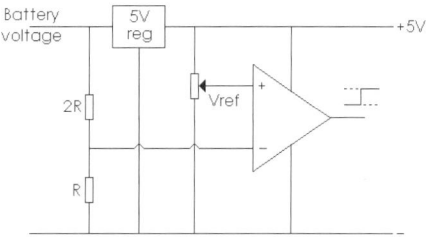

regulator with a value less than the expected lowest likely battery voltage. The power connections to the OP amps are not shown. This type of circuit works well with OP amps such as the LM324 that only require a positive and a ground supply.

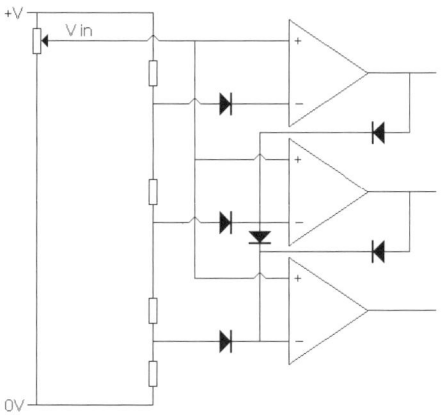

Fig.236 shows a dot output using a series of voltage comparators. The output from the higher stage, when switched on, is used via the steering diodes to put a + voltage on the – input of the preceding stages. This voltage is higher than the + input reference voltage and that stage will be switched off leaving only the highest stage on. If a long string of OP amps is used the voltage

Fig.237 shows a standard OP amp or voltage comparator being used for voltage level sensing. Because all power is being supplied by one battery it is necessary to use a regulator that has a voltage output below the lowest level the battery will drop to. Account must also be taken of the dropout voltage of the regulator. Because some OP amps do not like an input voltage higher than Vcc a voltage divider is used on the voltage being measured. In this example with the R/2R resistor network the voltage on the –input of the OP amp is one third of the main supply voltage. Therefore for a nominal 12V the voltage on the –input of the OP amp is 4V. If it was required that the output go positive to indicate that the voltage had fallen to 9V, the V ref is set one third of 9V i.e. 3V.

With a charged battery the –input is greater than the +input therefore the output of the OP amp remains low. When the main voltage falls, a point is reached when the –input is greater than the +input therefore the output of the OP amp goes high. This can be used to light an LED or some other warning device. An electrolytic capacitor fitted between ground and the R/2R resistor junction can give smoothing to the voltage being sensed and the unit will only react to steady state voltages not drops caused by current drops when for example a motor starts under heavy load.

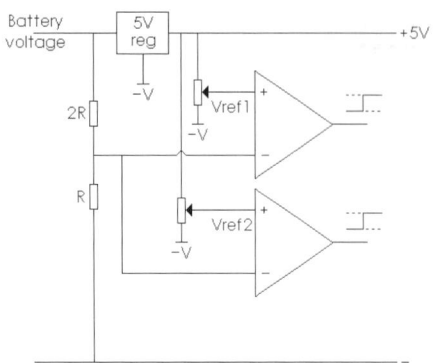

The circuit shown in **Fig.238** works in the same way as the previous circuit but two reference voltages are used. The first voltage level sensing can be used as previously to provide a warning. If the warning is ignored and the main voltage continues to fall, the second lower reference point can be used to initiate a shut down procedure. This shut down can be removal of power from the high current circuits or can be used to cause a No Volt Release unit (NVR) to drop out.

CHAPTER SEVENTEEN

Battery Current Monitors & Cut Outs

These circuits are described as battery current monitors and cut outs only because batteries are a finite power source. The circuits will work with the normal mains supplies described.

Battery current monitors and cut outs extend from simple metering to controlled shut down procedures in the event of excessive battery current.

For the purposes of the following circuits it is necessary to differentiate between current measuring and current monitoring. Current measuring is the reading of an actual current that may changes with the battery load. Current monitoring is the use of that current reading to trigger an external event. The event may be the lighting of an LED or it could be a warning or a shut down sequence.

The simplest form of voltage measurement is to use a moving iron current meter; these can be obtained with ranges into hundreds of A. Digital current meters are a better solution but tend only to be available up to about 10A. These are available as panel mounting units. With the increasing use of the digital multimeter it is now possible to buy a low-end version for below £5. This contains everything needed to produce, with a small amount of work, a reasonably accurate panel meter with a range up to 10A. Many measurement techniques use secondary effects. The simple moving coil voltmeter uses the current flowing through a coil to move a needle by magnetic attraction or repulsion. An OP amp can be used to measure the voltage drop across a resistor and heating or magnetic effects of the current flow can be used to produce an output.

The simplest example of the heating effect is the fuse. This is an insulated tube with a wire element that heats up with current flow. If sufficient current flows through the element it will heat up to melting point and cut off the current flow.

Auto reset breakers are an electronic version of the fuse, available up to about 10A. These reset automatically after a short period of time. Because of the danger of auto reconnection they should only be used with a NVR or in units where reconnection is not a danger.

Many motor failures occur not because of large over currents but because of long term small over currents. Direct temperature monitoring of the motor is the best solution but the heating effect can also be used to monitor long term over current. AC motor start contactors usually have cut outs that consist of a heating winding around but insulated from a bimetal strip. When the bimetal strip bends it causes the contacts to mechanically break. Normal current will not cause the bimetal strip to bend but excess current will. The heating and cooling process takes time therefore peak currents such as starting currents are 'averaged' out. A similar unit can be made for low voltage applications by using a low resistance power resistor with an electronic temperature sensor mounted on it.

Magnetic effects tend to be used for high currents because the voltage drop through even a low value resistor can be excessive. Experiment can produce devices based on the physical attraction of a magnet or a piece of iron to a coil. The coil will be of thick wire and very few turns to keep the resistance low. The device can be made to trip a micro switch and turn off the NVR when the current limit is reached.

Probably the best commercial high current unit is the Hall effect or the magneto resistive current sensor. This is relatively inexpensive. It is described in the sensor section. The unit can be made to give a direct display or the output can be used to trigger an event via a voltage comparator. Home made versions can be made from a ferrite ring and a Hall effect or a magnetoresistive sensor. They are useful for secondary switching units such as dust extraction but are slightly more difficult to calibrate for high current use. Ferrite is difficult to cut because of its hardness but can be cut with a diamond blade. Inexpensive wet electric rotary tile cutters can be used successfully but there is still a tendency to splinter on the edges. Diamond blades are available for miniature electric rotary tools. Safety precautions should be observed.

There are also available commercially a limited range of low voltage circuit breakers similar to the MCBs used in 'fuse boxes' or distribution panels for mains AC in the home.

Fig.239 shows the voltage drop across a low value resistor being used with an OP amp. Using standard range components the OP amp give a voltage gain of approximately 40. This allows a 0.1Ω resistor to indicate a current range of 0 to 2.5A with a voltage output of 0 to 10V. The problem with this type of circuit is the limited range of low value power resistors. These are commonly available to 0.1 ohms. It is possible to parallel these resistors to increase the current range but this is only practical for a few resistors. This means the measurement or reading of the very high potential currents from battery circuits becomes more impractical the higher the current range. It also means that the resistance of the resistor interconnect wiring can have a significant effect on the overall resistance. Practical home built high current sensors.

There are a large number of Hall effect sensors on the market, the original commonly available type dating from 1978 being the UGN3501/03 series. I have very recently seen circuits in hobby magazines advocating the use of the UGN3503 for current sensing applications. But the manufacturers state that the UGN3501/3 meet the basic requirements for contactless sensing but are extremely sensitive to temperature changes and mechanical stress. The UGN3501 is no longer manufactured and replacement higher specification sensors for this and the UGN3503 are the A3515/6 family. The A3516 will produce an output of 2.5mV/G compared to the 1.3mV/G of the UGN3503 with a Vcc of 5V.

Current flowing through a wire produces a magnetic field of about 6.9 gauss per amp. Therefore for currents between a few hundred and a few thousand amps can be read directly from the field produced by the wire. The distance from the centre of the wire to the sensor affects the reading.

The formulae $B \approx I/4\pi r$ or $I \approx 4\pi rB$ can be used for calculation.

Where:
    I = current in amps
    B = field strength in gauss
    r = distance from centre of wire to the sensor IC in inches

Example: a wire of radius 0.1ins, with a 0.1ins air gap (including insulation) with a current flow of 250A.

Using $B \approx I/4\pi r$ $B \approx 250/4\pi\, 0.2$ $B \approx 99$ gauss

Based on 2.5mV/G this approximates to 248mV

For currents up to about 120A it is better to use a magnetic field concentrator in the form of a ferrite toroid with a gap for the sensor.

The output from the toroid unit is $B \approx 6.9 G/A$ for a single wire through the toroid

This expands to $B \approx 6.9 nI$ gauss where n is the number of turns through the toroid.

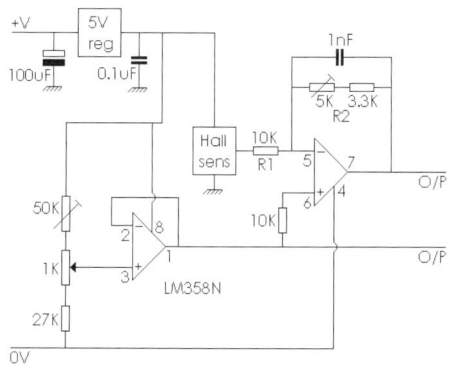

Fig.240 shows a circuit for a Hall effect current sensor unit. The no current flow output of the sensor is Vcc/2 i.e. 2.5V and the output per gauss is 2.5mV/G for the A3516 with a Vcc of 5V. If the Vcc increases to 5.5V, the no current flow output of the sensor is Vcc/2 i.e. 2.75V and the output per gauss is 2.75mV/G. Therefore a stable power supply is necessary for consistent results. The current requirements of the circuit are low therefore a standard 5v regulator can be used. The circuit is based on a double OP amp such as the LM358N. The sensor can produce a positive and a negative output relative to Vcc/2. Output saturation occurs at approximately Vcc/2 +/- 80% of Vcc/2 i.e. at 2.5 – 2V = 0.5V and 2.5 + 2V = 4.5V for a Vcc of 5V. The possible output therefore at 6.9G/A and 2.5mV/G this approximates to +/- 120A.

One section is used as a unity gain buffer that is set to give an output of Vcc/2 to provide a bias for the second OP amp and a zero level adjustment. The second OP amp is configured as an inverting amplifier and for use with the A3516 the amplification is set to approximately to 0.58. This gives an output of approximately 1mV per Amp.

The circuit can be set up practically using a secondary winding. This winding consists of a larger number of coils than the main wire the current is sensed from. If the winding consists of 20 turns through the toroid the current to give an output equivalent to 100A is 5A. This can be set up using a battery and a bulb. A car battery feeding a headlight bulb of 50 watts approximates to a current of 4A; the actual current can be read using a multimeter.

The full set up is as follows. Measure Vcc and with no current flowing through the sensor unit set the bias voltage from the unity gain buffer to be Vcc/2. The 50K variable resistor and the 1K potentiometer allow a coarse and a fine setting. Set the feedback potentiometer on the inverting amplifier to about half way. Turn on the test current through the toroid and measure the current with a multimeter. Adjust the output between the amplifier and the bias voltage with another meter to be the measured current multiplied by the number of turns. E.g. if the measured current was 4.2A with a 20 turn set up coil then the output should read 4.2 x 20 = 84mV. At 1mV/A this is equivalent to a single wire output of 84A.

The coil can be left in place for future calibration if the ends are insulated.

The accuracy of the setting is dependent on the accuracy of the meters used for the calibration. For most practical purposes standard meters are satisfactory.

Because the current measurement does not rely on a voltage drop in the sensor the heat produced is minimal.

This circuit can be modified to different ranges with different units. For all practical purposes the gain of the amplifier stage can be regarded as – R2/R1. The minus sign is because the amplifier is inverting. The circuit will work with toroid core units or single wires.

For low current a number of turns can be put through the toroid in the same way as the calibration coil.

Although I cannot find mention by the manufacturer, it seems possible from practical experiments to 'desensitise' the sensor by making the gap in the toroid substantially larger than the thickness of the toroid and fixing the sensor in the middle or to one side of the gap. No quantitative results have been tabulated so this is a matter of trial and error.

The BM301P display microprocessor

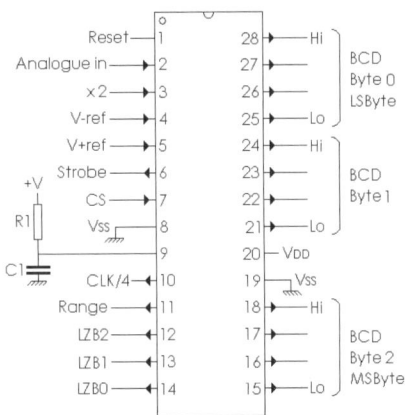

Fig.241 shows the BM301P analogue input device with 3 byte BCD output. This IC accepts analogue inputs from a range of sensors. The IC can be used to directly display inputs from linear temperature sensors, voltage readings to a maximum of +50V and various current sensors. The output of certain devices may require attenuation or amplification see typical applications.

It converts an analogue input to a 3 byte BCD output. The input is decoded to a 10 bit digital coding. The display reads 0 to 9 by single step. A times 2 multiplier is available that allows the display to read 0 to 1998 by steps of 2.

The input can be referenced to a positive reference and a minus reference offset is available.

It is feasible that the device could be made to read a wide range of other inputs provided that a suitable linear input sensor. Examples could be linear high temperature thermocouples and liquid level measurement.

Display drivers

The device will drive many types of LED and LCD display drivers and decoders e.g. the 4511. It will also drive inputs to other types of devices that require a BCD input within the voltage range of the device. The device will drive many types of LED and LCD display drivers and decoders e.g. the 4511. It will also drive inputs to other types of devices that require a BCD input within the voltage range of the device. See specific data information of other devices.

Current meter to 999 amps

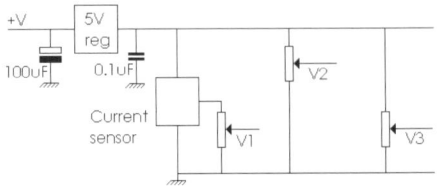

Fig.242 shows how to use the device for a theoretical current sensor working on +5V with a Vcc/2 offset and a 2 mV per amp output.
Set potentiometer on the output of the current sensor to divide by 2. This gives an output V1 of 1mV per °C. Connect to the analogue input.
Set V3 output to Vcc/2 i.e. 2.5V and connect to V-ref. This is the offset. Set V2 output to 1.024V + 2.5V i.e. 3.524V and connect to V+ ref. This is equivalent to an input range of 1mV per °C with Vcc/2 offset.
Connect X2 display range multiplier to ground.

Volt meter to +50V

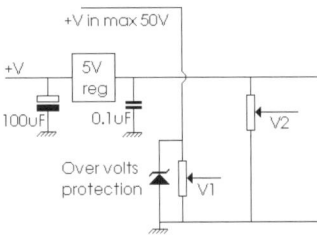

Fig.243 shows how to use the device as a low voltage indicator.
Set the potentiometer V1 on the input voltage to divide by 10. This gives an output V1 of 0 to 5V for an input of 0 to 50V. Connect to the analogue input.
Set V2 output to 1.024V and connect to V+ ref. This is equivalent to an input range of 1mV per V.

Pin designation		
Pin	Pin name	Description
1	Reset	Resets when grounded, tie to + rail if not used.
2	Analogue in	Analogue input to max of Vcc rail.
3	X 2	Display range multiplier, ground for x 1, to + rail for x 2.
4	V - ref	Sets lower level of analogue signal. Can be used to cancel out offset of input amplifiers or current sensors.
5	V + ref	Used to set range of analogue input.
6	Strobe	Goes low during data to output transfer then returns high.
7	CS	Chip select, active positive, tie to + rail if not used.
8	V_{SS}	Zero or ground connection
9	OSC	RC oscillator input.
10	CLK/4	Oscillator test. Output is a quarter of the RC oscillator input.
11	Range	Positive output if x 2 range selected. Can drive a low power (10mA max recommended) LED directly.
12	LZB 2	Leading zero blanking (+V out) for BCD byte 2.
13	LZB 1	Leading zero blanking (+V out) for BCD byte 1.
14	LZB 0	Leading zero blanking (+V out) for BCD byte 0.
15	BCD Byte 2 - 1	Byte 2 LSB binary value 1
16	BCD Byte 2 - 2	Byte 2 LSB binary value 2
17	BCD Byte 2 - 4	Byte 2 LSB binary value 4
18	BCD Byte 2 - 8	Byte 2 LSB binary value 8
19	V_{DD}	Positive rail connection *1
20	V_{SS}	Zero or ground connection
21	BCD Byte 1 - 1	Byte 1 LSB binary value 1
22	BCD Byte 1 - 2	Byte 1 LSB binary value 2
23	BCD Byte 1 - 4	Byte 1 LSB binary value 4
26	BCD Byte 1 - 8	Byte 1 LSB binary value 8
25	BCD Byte 0 - 1	Byte 0 LSB binary value 1
26	BCD Byte 0 - 2	Byte 0 LSB binary value 2
27	BCD Byte 0 - 4	Byte 0 LSB binary value 4
28	BCD Byte 0 - 8	Byte 0 LSB binary value 8

*1 *The positive voltage range is +3V to +5.5V, but it is recommended that a regulated +5V power source is used. Inaccuracies will occur if variations between the IC supply and the sensor or input supply occur.*

Maximum analogue input is V_{DD}. Attenuation must be used with inputs exceeding this value. Over voltage input protection is recommended because the device may be damaged with voltage exceeding V_{DD}.

This gives a display range of 0 to 500. Either use only two BCD displays or use the display decimal point to produce an indication of 0.0 to 50.0V i.e. a resolution of 0.1V.
Connect V- ref and X2 display range multiplier to ground.

Temperature reading using linear temperature sensor

Fig.244 shows how to use the device as a temperature indicator.
Devices such as the LM35 produce a linear output up to 110°C.

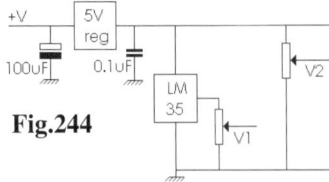

Fig.244

The output for this particular device is 10mV per degree C. This equates to 0V to 1.1V for a temperature range of 0°C to 110°C.
Set potentiometer on the output of the LM35 to divide by 10. This gives an output V1 of 1mV per °C. Connect to the analogue input.
Set V2 output to 1.024V and connect to V+ ref. This is equivalent to an input range of 1mV per °C.
Connect V- ref and X2 display range multiplier to ground.

Other possible applications

These include applications that have a linear DC output sensor that requires a BCD output or display. Attenuation or amplification may be required to bring the output to a suitable range. Possible applications include high temperature reading using a suitable thermocouple, or liquid level measurement.

Battery reverse voltage connection prevention
Reverse polarity with a low impedance battery can result in total 'melt down' of components in electronic circuits. There are two basic ways to prevent the effect of battery reversal.

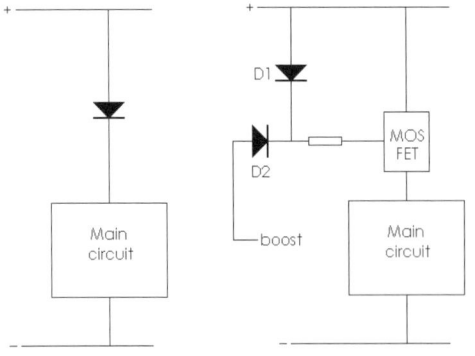

Fig.245 shows the two basic ways. The first is to use a diode as a block if the incorrect polarity occurs. The problem with this circuit is that the diode will need to conduct the full circuit current. Combine this with the voltage drop across the diode and a high dissipation occurs.

The second circuit uses a MOSFET as the blocking device. MOSFETs tend to be less expensive than diodes for the equivalent current rating. More important they have a lower on resistance and therefore dissipate less heat. When power is turned, on a voltage will appear on the gate of the MOSFET via D1 if the polarity is correct. It will not be sufficient to turn the MOSFET fully on but there is usually enough current to drive high side or boost drivers within the main circuit. When the boost drivers begin to charge up, voltage will be supplied to the gate of the MOSFET via D2. This circuit can be temperamental at switch on and the MOSFET needs to be chosen carefully to have the required current capacity with the lowest on resistance and the lowest practical gate start of turn on voltage.

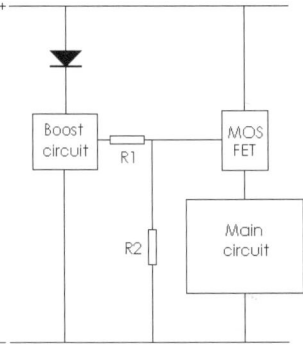

Fig.246 is a combination of the two previous circuits. It is reliable and easily implemented without the need to be over selective with the components used. A diode supplies the current for a separate boost circuit that turns on the MOSFET supplying the main circuit. The boost circuit can be a low power 555 type. The diode supplying the boost circuit only needs to supply the current for the boost circuit. This is in the order of a few milliamps and therefore dissipation in the diode is negligible. The diode can be for example a 1N400X type.

NVRs and interlocks

This is such a simple device that I can see no reason why all the projects described in this book should not make use of NVR units for mains and low voltage, high current supplies. The NVR – no volts release unit is an on switch that in the result of a power loss will 'drop out'. When power returns the on switch needs to be pressed again before the unit will start.

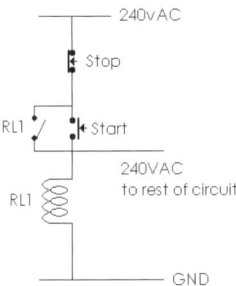

Fig.247 shows a mains circuit that can replace a mains switch for a power supply. The start switch applies power to the unit and to the relay RL1. When RL1 closes the relay contact

A commercially produced NVR switch.

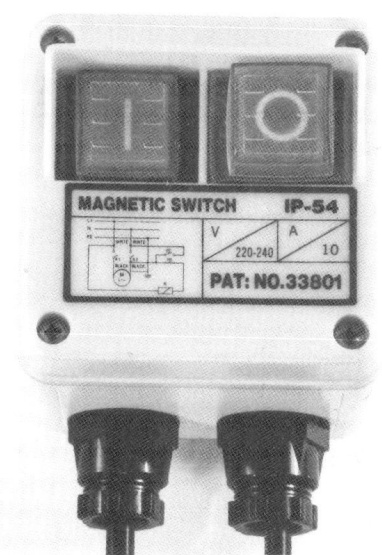

maintains the circuit to the relay when the start switch is released. When the stop switch is pressed or if there is a mains power loss the relay drops out of circuit. Because it is the switches not the relay contacts in this circuit switch the load the relay rating can be as for a resistive load.

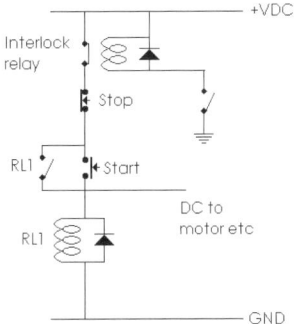

Fig.248 shows a low voltage circuit to replace the standard on/ off switch. This can be used with DC power supplies or high capacity batteries. The interlock relay switch is deliberately in the ground side of the circuit because this is the wiring that is at risk of chafing etc. Any wire to ground short in this section will result in the interlock relay operating and not a short circuit of the high current power line.

Interlocks are devices that prevent a series of potentially dangerous or unpleasant situations occurring. The most common forms are in machinery when for instance opening an access panel prevents the motor being switched on. Another example is on a washing machine when a lock prevents the door being opened if the machine is full of water.

Interlocks are easily built into a circuit or as a mechanical device e.g. the positioning of the on/ off and forward/ reverse switches on a tool in such a position that when the on/ off switch was in the on position the forward/ reverse switch could not be operated.

Interlocks are also easily fitted within the control electrics/ electronics to prevent an event happening in relation to another event. Interlock functions can be used with AND and NAND type

Left: Inside of a commercially produced NVR switch. The glass phial is an over current cut out. Above: Home made NVR unit fitted to drill press.

gates or can be used to 'drop out' the NVR switch. Whilst interlocks are an important safety feature of many designs they should not be totally relied on from a safety point of view. Even safety devices can fail. Power should be removed completely from a unit if modifications or adjustments are being made.

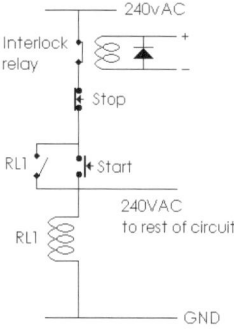

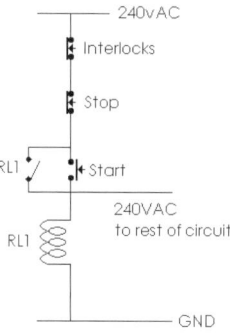

Fig.249 shows a NVR circuit with interlocks in the mains line.

Fig.250 shows a NVR circuit with a low voltage interlock relay in the mains line. The relay is operated from the DC power supply. This type of circuit is not as reliable as direct switching because failure of the low voltage relay circuit will prevent the interlock relay working.

CHAPTER EIGHTEEN

Ancilliary Test & Driver Modules

This is a collection of electronic blocks that can be used as test units or incorporated into other circuits or used in stand-alone mode.

Stepper motor pulse driver

This circuit can be used both for testing and as the basis of a manual control system for stepper motors.
The circuit is a gated astable built around the common low power 555.
The circuit uses steering diodes in the timing circuit so that charge time and discharge time and therefore pulse width and pulse period are independent.
R3 sets the minimum off time and limits the discharge current from the capacitor. Pin 4 is the gating pin and is normally taken direct to positive to allow an output on Pin 3. Taking pin 4 to ground stops the 555 giving an output. If this part of the circuit were to be inverted then a push button switch could be used to allow stepping.
With a standard astable circuit the frequency is $F = 1.44/(Rc + 2Rd) \times C$ where Rc is the resistance of the charging resistor and Rd is the resistance in the discharge circuit. This is because the charging uses both the charging and discharge resistance to charge the capacitor and the discharge circuit uses only the discharge resistance. C is in Farads.

With steering diodes the charge and discharge circuits use only their own resistors therefore the frequency is
$F = 1.44/(Rc + Rd) \times C$
Suitable values for this circuit are
R1 = 4.7KΩ, R2 = 470KΩ potentiometer, R3 = 10KΩ, C = 1μF
This will give a range of about 3Hz to 100Hz
For slow speed stepping changing C to 10μF will give a range of about 0.3Hz to 10Hz
Polarised capacitors should be fitted with the negative end to ground.
Suitable steering diodes are 1N4148.

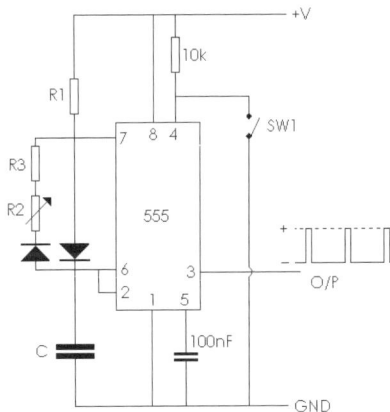

Fig.251 shows the basic circuit of the pulse unit

155

with the gating normally on.

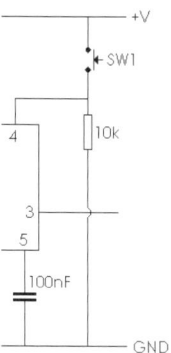

Fig.252 shows the gating section of the pulse unit that allows pulses out when a push button is pressed.

PWM pulse driver

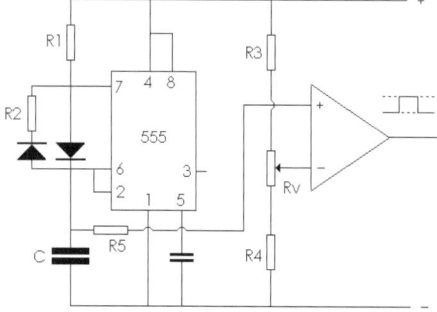

The circuit shown in **Fig.253** can be used both for testing and as the basis of a manual control system for DC motors.

The circuit is a gated astable built around the common low power 555.

The circuit uses steering diodes in the timing circuit so that charge time and discharge time and therefore pulse width and pulse period are independent. The output of the astable is not used. The input the OP amp is taken from the charging ramp of the capacitor. When the voltage on the ramp exceeds the voltage on the – input of the OP amp the OP amp switches on. The voltage swing on C is 1/3V to 2/3 V. The resistors R3 and R4 are to set the range of Rv to slightly exceed 1/3V to 2/3V therefore ensuring

the output range is 0 to 100%. R5 is to reduce the loading on the capacitor.
Suitable values for this circuit are:
R1 = 82KΩ, R2 = 470Ω, Rv = 10kΩ potentiometer, R3=R4=8.2KΩ, R5=1kΩ, C=1µF
This will give a frequency of about 18KHz.
This circuit has a number of limitations. The charging rate of the capacitor is exponential not linear therefore the output does not change in a linear fashion. The type of OP amp used has a major effect on the circuit. If the OP amp has a slow slew rate the switching is not square but is a slope. The slew rate is the change in output voltage with time and is expressed as V/mS.

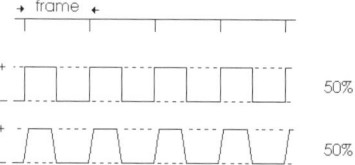

Fig.254 shows the PWM switching. The slope means that the output driver is not switched quickly between the minimum dissipation states off full on and full off and can produce increased heat dissipation in the driver.
Also some OP amps will work reasonably in comparator mode whilst others are too 'sensitive' to voltage levels and may jump from about 75% to 100% very quickly making control in this range very difficult. This can be overcome by setting the OP amp to a level that is stable.
Suitable types of OP amp for this circuit are LM301 and LM351.

Pulse width servo driver

Fig.255 shows a circuit for use with commercial servos. This consists of an astable running at the frame time driving a monostable. This frame time is usually in the order of 20mS. This is a frequency of 50Hz. The astable supplies a short negative going pulse that triggers the monostable via a capacitor. The capacitor ensures that the trigger pulse is shorter than the output pulse. The standard 555 monostable can

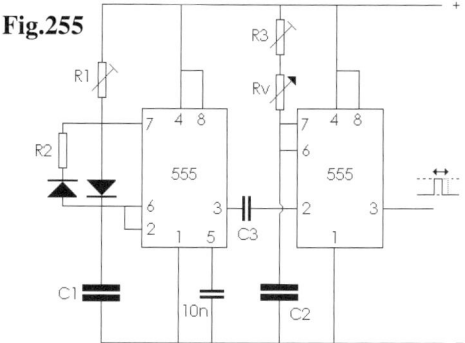
Fig.255

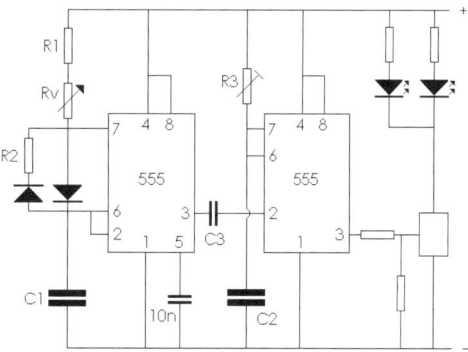

only put its output pulse low when the input has returned high. The output of the monostable is a positive going pulse that is a minimum of 1mS and variable over the range of 1 to 2mS. If a joystick were to be used the variable resistors generally only cover part of their full range and this needs to be taken in to account when choosing values of Rv.

Suitable values for this circuit are R1 = 100KΩ potentiometer, R2 = 1kΩ, R3 = 10kΩ potentiometer, Rv = 10kΩ potentiometer, C1 = 0.47µF, C2 = 0.1µF, C3 = 10nF

R3 allows the setting of the shortest time range. Rv gives slightly over 1ms variation.

Simple stroboscope

A stroboscope is a short pulse of light at a set time frame. Originally these used high voltage strobe tubes that discharged a high voltage pulse through a gas. The result was an intense flash of light. With modern high output LEDs it is simple to make a low voltage version of this.
The shorter the pulse the less distance the moving object has travelled during the on time of the light hence the clearer will be the perceived image. The strobe can be used to make rotating objects appear stationary or can be used with timing marks or the timing discs discussed earlier to measure rotational speed.
LEDs can be driven at many times their power ratings for short pulses. Care must be taken that the frame time does not become so short that the overall power capability is exceeded. High power white LEDs are now commonplace and are used in many areas of low voltage lighting.

Fig.256 shows a monostable being driven from an astable providing the timeframe this gives a short output pulse each time an input pulse is received from the astable.

Suitable values for the astable circuit are Rv = 100KΩ potentiometer, R1 = 1kΩ, R2 = 1kΩ, potentiometer, C1 = 1µF

This gives a frequency of about 14Hz to 720Hz. If C1 is changed to 10µF the frequency range will be about 1.4Hz to 72Hz.

To put this in context 3000RPM is 50Hz and 300RPM is 5Hz.

Suitable values for the monostable circuit are: R3 = 100kΩ, C2 = 0.1µF, C3 = 10nF

These give a maximum of about 11mS. The lower limit will be affected by the pulse width on C3. At 3000RPM 10mS relates to 180° of rotation. 0.1mS relates to 1.8° of rotation at the same speed. Therefore in practical terms the aim would be for about 0.3mS as a start point. This is about 5° of rotation.

The switch on time of a high power LED is typically less than 100 nanoseconds.

It will be necessary to calibrate the astable circuit and this could be achieved by matching to a series of timing discs or by direct reading of the astable output on an oscilloscope or frequency meter.

Fig.257 shows the LEDs being driven from a monostable giving a short output pulse each time an input pulse is received.

Suitable values for this circuit are as previously: R1 = 100KΩ potentiometer, C1 = 0.1µF, C2 = 10nF

The output driver can be any type that will carry

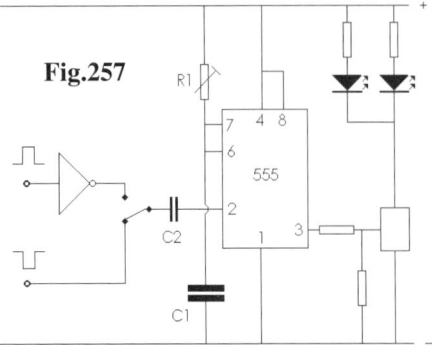

Fig.257

the current for the LEDs. It is exactly the same set up as used for motor driving but without the freewheeling diodes. The inverter attached to C2 allows the strobe to be triggered from external negative or positive going pulses.

Multiple LEDs can be used. See section on LEDs for calculation of resistors.

Sequence driver

It is sometime necessary to have a sequence of events occur from one trigger event.

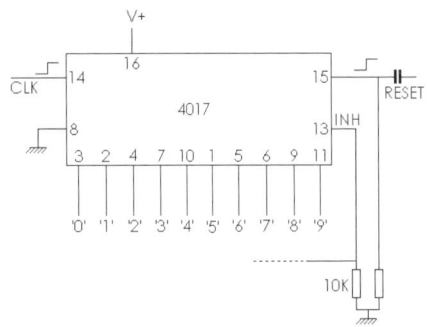

Fig.258 shows a simple sequence driver. The circuit is based on the 4017 counter. The IC is clocked from an external clock source that can be for example a 555 or 4047 astable. When a positive going pulse is applied to the reset pin the counter sets '0' as a positive output. The circuit counts up with each clock input. If one of the outputs is tied to the clock inhibit pin the count stops. A positive going pulse applied to the reset pin starts the sequence again. It is therefore possible to produce a sequential pulse output that cycles through the sequence at each

input pulse. The first pulse is of indeterminate length because the clock signal is not synchronised with the trigger pulse. Simple gating to synchronise or directly inhibiting the clock could solve this problem. Using intermediate monostables can allow this circuit to provide different pulse lengths for each pulse. This circuit can also provide a continuously running sequence by taking the clock inhibit to positive and using output after the last output required to put a pulse on the reset.

Calculator up/ down counter

A practical example of the use of the sequence driver is shown with the pocket calculator being put to a secondary use. The simple pocket calculator contains a multi digit display, keyboard and often a small memory for a price that is a fraction of buying the individual components.

This unit is based on the fact that a sequence of key presses can give an up or a down count sequence.

If we first look at the simplest set up that is the up counter. Pressing the (+) then the (1) then the (=) key will result in the display incrementing by one. The down count operation can be used by initially putting a number into the calculator manually. Pressing the (-) then the (1) then the (=) key will result in the display decrementing by one.

If the keys could be pressed automatically in sequence from one change of logic then the calculator would be acting as an event counter. Most calculators require that one key is released before the next is pressed therefore pulse shortening techniques may be required.

Fast operation will cause the display to change rapidly through all the steps but some scientific calculators put the event steps to a separate part of the display and the count result is always visible. This is the basis of an automatic pulse sequence.

A counter based on a pocket calculator using propagation delay

The simple pocket calculator contains a multi digit display, keyboard and often a small memory for a price that is a fraction of buying the individual components.

This unit is based on the fact that a sequence of key presses can give an up or a down count sequence.

If we first look at the simplest set up that is the up counter. Pressing the (+) then the (1) then the (=) key will result in the display incrementing by one. The down count operation can be used by initially putting a number into the calculator manually. Pressing the (-) then the (1) then the (=) key will result in the display decrementing by one.

If the keys could be pressed automatically in sequence from one change of logic then the calculator would be acting as an event counter. Most calculators require that one key is released before the next is pressed therefore pulse shortening techniques may be required.

Fast operation will cause the display to change rapidly through all the steps but some scientific calculators put the event steps to a separate part of the display and the count result is always visible.

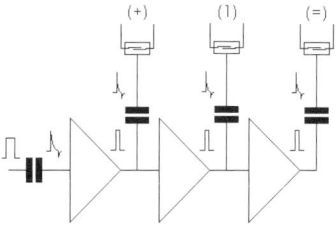

Fig.259 shows the basis of an automatic pulse sequence. The gates are used to produce a very short pulse to a bilateral switch such as the 4066 shorting out the required keys in sequence. Similar results can of course be obtained by using a counter such as the 4017 as described previously.

Mains clocking

Fig.260 shows a means of obtaining accurate signals at 50Hz and 100Hz from a 50Hz supply and at 60Hz and 120Hz from a 60Hz supply. The mains frequency is accurate enough to keep synchronous clocks running at the correct time

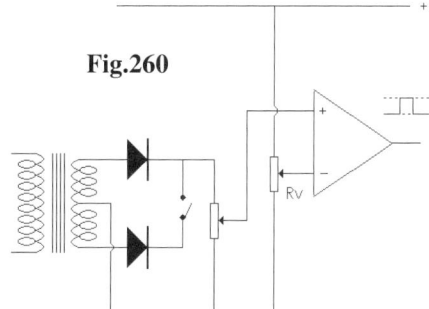

Fig.260

for long periods of time and is therefore accurate enough for most practical purposes. The output can be used to trigger the external driven strobe circuit described previously.

The diodes rectify the positive going and negative going half waves of the AC voltage to produce positive going half sine waves. With the switch open every second peak is allowed through this is a 50Hz waveform. With the switch closed every peak is allowed through this is a 100Hz waveform. The potentiometer connected to the diodes is to reduce the voltage to a suitable level for feeding to the OP amp. A good starting point is ground to peak about half the OP amp logic level. Rv controls the level at which the OP amp turns on and off. The result is a waveform with square edges as opposed to the original sine form. The setting of Rv also affects the width of the pulse.

The circuit does not need a dedicated transformer. Depending on the design it is often possible to add the necessary small winding to the existing mains transformer or add a tap to the power supply circuit as in **Fig.261**.

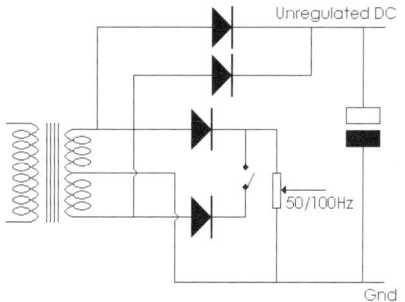

Ancilliary Test & Driver Modules

CHAPTER NINETEEN

Basic Electronic Building Blocks

This is not an attempt to teach electronics in depth but to show a number of components that if treated as a building block can be used effectively with the minimum of electronic knowledge.

There are two types of logic ICs available generally. These are 74 series and 4000 series. The 74 series is subdivided into 74LS, 74HC and 74HCT. In general terms the 74 series has a larger range of product covering a larger range of drive compatibilities. The 74 series requires a power supply of 5.1V. The 4000 series CMOS contains a range of all the ICs likely to be used with the type of circuits described. The 4000 series can be used with power supplies up to 15V. They also have a wide temperature range and high voltage protection. Therefore the gates etc described are based on the 4000 series. There are many options and pin outs are shown in many electronic suppliers' catalogues.

Basic logic ICs

The following is a description of basic logic units. The simplest form is described but the package may contain a number of the units or the individual units may have multiple inputs e.g. the and/nand gate is typically available with 2, 3, 4 or 8 inputs to 1 output.

Reference may be made to a Schmitt trigger input. This input is available on many logic gates and is often used as a means of obtaining a fast transition from off to on and on to off from inputs that are relatively slow changing. Many circuits that are triggered by transition as opposed to actual level need relatively fast transition times. The Schmitt trigger remains in a stable state until the trigger level is reached then switches quickly.

Logic is often expressed in terms of true and false instead of voltage levels. This is because some of the different IC families switch at different voltage levels.

References may be made to Q and not Q outputs. The 'not' term is often shown by a bar over the term. The Q term is the output and the not Q is the inverted Q term.

Many of the ICs may contain multiple gate types. It is good practice to connect the inputs of any unused gates to one of the rails. This prevents spurious switching of the associated output. This spurious switching can, depending on the design of the circuit, cause interference to other parts of the circuit. This is often because of the stray capacitance that will always occur within any circuit.

Digital IC family traits

The following is a brief description of the main series of digital ICs commonly available. Manufacturers date sheets are the best source of specific information and ratings of each particular IC. Drive capability varies, as does compatibility between families. Mixing IC families may need pull up or pull down resistors on

inputs to allow drive between different types of IC.
Where references are made to a particular IC in the following block diagrams there are a number of alternatives depending on the frequency of operation etc. E.g. the 4060 has versions in the 74HC and 74HCT families offering much higher frequency of clock input.

74 TTL series

The original digital ICs were mainly of the 74 series TTL ICs. These used a 5V power line and were relatively fast but consumed a lot of power. Other families of IC have now mainly superseded them.

4000 CMOS series

The 4000 series followed these could accept power supplies typically up to 15V and became popular because they could be used in many applications without the need for a regulated power supply. They also used much less power than the 74 series TTL. They are still easily available. The 4000 series is not particularly fast and the speed is dependant on the supply voltage. They may typically have a maximum operating frequency of 3MHz at 5V, a maximum frequency of 8MHz at 10V and a maximum frequency of 10MHz at 15V.
74LS low power Schottky bipolar series
The 74LS Schottky bipolar series is compatible with TTL but uses much less power and offers higher operating speeds. Typical operating frequencies are in the order of 50MHz.

74HC CMOS series

The 74HC CMOS series gives the low power and high noise immunity of CMOS at the speed and performance of 74LS series. Typical operating frequencies are in the order of 50MHz.

74HCT CMOS series

The 74HCT CMOS series gives the low power and high noise immunity of CMOS at the speed and performance of 74LS series and is also TTL compatible. Typical operating frequencies are in

the order of 50MHz.
The above are the most common and readily available IC families with the exception of the original 74 series that now tends to be available less readily and tends to be used mainly in the repair of older OEM equipment.
There are also a number of other families that have specific characteristics i.e. 74AC, 74ACT, 74ALS, 74F and 74LVC.

Device symbols and truth tables

Buffer

Fig.262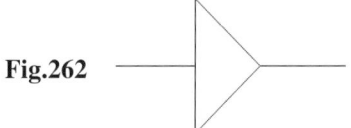

Buffer	
I/P	O/P
0	0
1	1

This is a circuit where the output is the same logic state as the input. It is often used to increase the 'fan out' or decrease the loading on a circuit that may be affected by a variable load. Fan out is the measure of the gate inputs that can be driven from an output.
Examples are the 4050. The symbol is shown in **Fig.262**.

Inverter/ inverting buffer

Fig.263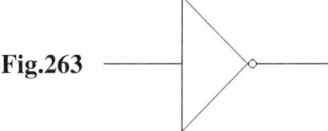

Inverter	
I/P	O/P
0	1
1	0

Basic Electronic Building Blocks

This is a circuit where the output is the opposite logic state to the input. Inverters are available just as inverters or combined with buffers. Examples are the 4049. The symbol is shown in **Fig.263**.

And gate

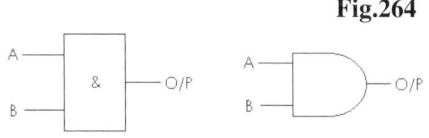

Fig.264

AND gate		
A	B	O/P
0	0	0
1	0	0
0	1	0
1	1	1

The and gate is a circuit where all of the inputs must be true to give a true output.
Examples are the 4073, 4081 and 4082. The symbols are shown in **Fig.264**.

Nand gate

Fig.265

NAND gate		
A	B	O/P
0	0	1
1	0	1
0	1	1
1	1	0

The nand gate is a circuit where all of the inputs must be true to give a false output. It is an and gate with an inverter on the output.
Examples are the 4011, 4012, 4023, 4025 and 4068. The symbols are shown in **Fig.265**.

Or gate

Fig.266

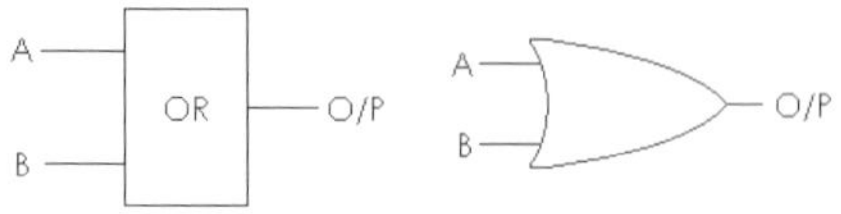

OR gate		
A	B	O/P
0	0	0
1	0	1
0	1	1
1	1	1

The or gate produces a true output when any or its inputs are true.
Examples are the 4071, 4072 and 4075. The symbols are shown in **Fig.266**.

Nor gate

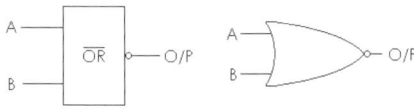

Fig.267

NOR gate		
A	B	O/P
0	0	1
1	0	0
0	1	0
1	1	0

The nor gate produces a false output when any or its inputs are true. It is an or gate with an inverter on the output.
Examples are the 4001, 4002 and 4078. The symbols are shown in **Fig.267**.

Exclusive or gate

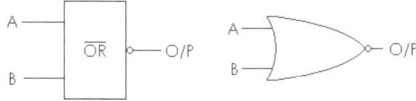

Fig.268

162 Electromechanical Building Blocks

XOR gate		
A	B	O/P
0	0	0
1	0	1
0	1	1
1	1	0

The exclusive or gate sometimes referred to as XOR gate is an or gate that produces a true output when only one of its inputs are true. Examples are the 4070. The symbols are shown in **Fig.268**.

Exclusive nor gate

Fig.269

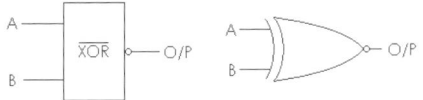

XNOR gate		
A	B	O/P
0	0	1
1	0	0
0	1	0
1	1	1

The exclusive nor gate sometimes referred to as XNOR gate is a nor gate that produces a false output when only one of its inputs are true. Examples are the 4077. The symbols are shown in **Fig.269**.

Set/ reset latches

A set/ reset or SR latch is a device that has two stable states activated by a pulse on the relative input. Two versions are available those that trigger on positive inputs and those that trigger on negative inputs. Examples are the 4043 and

4044. The symbols are shown in **Fig.270**.

Dividers

These are ICs that divide the incoming clock signal or pulses to produce a divided output. Versions are available that divide in multiples of two and others can divide by five or ten. One version is available that has sequential outputs of zero to nine therefore odd numbers can be divided.
Examples are the 4017, 4040, 4060, 4018, 4022, 4024, 4010 and 4020.

Electronic switches

These are devices that are turned on by a positive input. The device has both of the ends of the switch output accessible and these can be attached to either rail or between logic level circuits. The output unlike a mechanical switch has a high resistance typically in the order of 50W. This may not matter for logic circuits but must be taken into account if this device is used for quantitative inputs such as R – 2R – 4R drivers.
Examples are the 4016 and 4066.

Op amps and comparators

Op amps are devices that have a plus and a minus input and an output. The output is relative to the inputs and the amount of feedback. The comparator gives a true output when the plus input is a higher voltage than the minus input. Comparators are available has dedicated units or can be configured from OpA.
Examples are the TL08X series, LM324 and LM339.

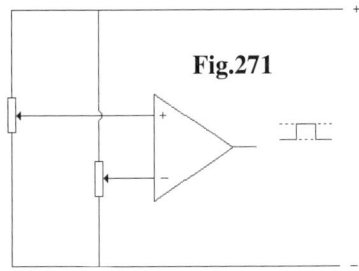

Fig.271

Basic Electronic Building Blocks

The easiest versions to use are those not requiring a negative voltage rail, i.e. they only require a positive and ground rail.
Fig.271 shows a typical comparator set up.

Astable

These are devices with two quasi-stable states. They can be built from discrete components or from dedicated ICs. They generally work on the basis of a capacitor charging through a resistor until a trigger level is reached and the state then changes whilst the capacitor discharge to a second trigger level. They are used as the basis of many simple clocks and repetitive triggering devices. Some devices can use polarised capacitors hence relatively long clock cycles are available. Some astables can be gated i.e. the output only appears when the gate is at a specific logic state.
Examples are the 555 and 4047.

Monostable

These are devices with one stable and one quasi-stable state. They can be built from discrete components or from dedicated ICs. They generally work on the basis of when triggered by an input a capacitor charges through a resistor until a trigger level is reached and the state then returns to the stable state. Some devices can use polarised capacitors hence relatively long cycles are available. Versions are available that can be triggered on positive going pulses, negative going pulses or can be retriggered.
Examples are the 555, 4047 and 4098.

Counting circuits

These are circuits that keep a count of the input clock pulses. The output is usually in the form of binary, hexadecimal or binary coded decimal. The ICs can usually be cascaded to give a multi number output.
Examples are the 4029.

Counting and display circuits

These are circuits that keep a count of the input clock pulses. The output can drive a 7 segment LED. Some of these ICs move each count directly to the display whilst others hold the count in a buffer until a control term allows the buffer to be moved to the display. This buffer type of circuit is useful for speed measurement etc. The ICs can usually be cascaded to give a multi number output
Examples are the 4043, 4026, 4511, 40110.

Using microprocessor ICs

Microprocessors are ICs containing many gates etc in one package. They are different to normal logic ICs in that they also contain a software or firmware program. This program may be mask written i.e. written as part of the production process or loaded onto a standard empty processor as a burnt in permanent load or into flash memory that can be written over. Most microprocessors have security measures that prevent the software being read at code level once the software is loaded. This is usually by fusing the software code read and load circuit. With flash devices attempts at reading the software causes the device to wipe the program from the memory.

Microprocessors are digital devices and need a clock to operate. They are capable of simulating most standard logic devices with ease and often with greater accuracy. Counts are made against an accurate clock not against the charge rate of a capacitor. Some devices have internal clocks but the majority require an external crystal or oscillator. Clock speeds can be many megahertz and this means that the speed of other ICs used in association must be capable of running with the speed relevant to the section of circuit they are used in.

Most microprocessors are task orientated i.e. the software was developed for one task only and the functionality is not transferable.

The microprocessors used in this book are for specific tasks but are general usage. This means they can be used for example in the case of the motor driver can be used by a very wide range of motors that is only limited by the driver that the microprocessor controls.

There are a few items that have been described in principle but no circuit is given for discrete components e.g. the proportional servo driver. This is because the design would use many ICs and analogue techniques that at the best would be a compromise of cost and reliability. The microprocessor uses one control device and the same drivers that would be needed for the analogue version. The accuracy is based on simple addition and subtraction not on analogue summing and amplification.

The devices used are also 'type' devices. This means they are likely to be available in the form used or in upgraded pin compatible form for many years to come.

Clocks, crystals and oscillators

Low speed clocks or clocks where extreme accuracy is not the prime consideration can be built from standard logic devices such as the astable.

High-speed clocks generally use some sort of resonating device. A limited range of devices is available as ceramic resonators and often have the load capacitors built in.

Most clock devices are built around the quartz crystal. This is available in a very wide range of frequencies. Microprocessors and a few counter ICs can accept direct connection to a crystal and load capacitors to create oscillation. The load capacitors are part of the oscillation circuit and different values are needed for different crystal frequencies and types. The information is available on manufacturers data sheets or in catalogues selling crystals.

Multi clocking

When multiple ICs are used that require clocking it is better that the clock source is from a single oscillator to ensure that devices remain in synchronisation. Most crystals are stable but they do have a temperature stability that is typically in the order of thirty parts per million. This is meaningless for most circuits but at the speed processors run at this can be significant if processors are data connected.

Some processors have an output pin that allows one device to act as the oscillator and the clock can be 'daisy chained' to all the other devices. Failing this output a quartz oscillator can be used. This is a device that is in effect a stand-alone crystal. The device is connected to a power supply and produces a clock output without the need to be connected to the microprocessor. The load capacitors are usually built in and the devices have a fan out typically of ten.

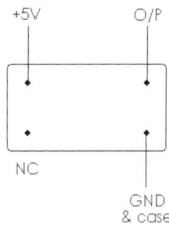

Fig.273 shows the layout of a typical quartz oscillator from the top.

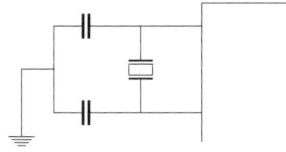

Fig.272 shows a typical crystal and load capacitor connection to a device.

CHAPTER TWENTY

Practical & Cost Effective Building

Cost is usually a factor in any building project particularly if the design has commercial implications. For the amateur builder ease of building and reliability are probably more important than a small extra cost. If a standard design unit can be used in different applications then building can be simplified and repairs undertaken easily and quickly.

Discrete components

With drives of a few amps the cost of a driver MOSFET with an individual heat sink is probably greater than the cost of using a higher powered MOSFET that does not require a heatsink. MOSFETS can be run in parallel to increase power and sometimes this is cost effective if buying a component in a reasonable quantity means that you can obtain a price break. Price breaks often start at five devices with 25 devices bought often giving a substantial percentage saving.

A quantity of 25 devices may sound a lot but a three-axis drive unit using 'H' bridge bipolar stepper motors will require 24 devices just for the motor drives.

If MOSFETs are only bought as single items or for a special application then it is better to buy a larger power device. Based on a suppliers catalogue a 169A MOSFET costs about 50% extra than a 62A MOSFET with a 210A MOSFET costing double that of the 62A MOSFET. These percentage increases in cost are not in proportion to the increase in power handling. The 210A MOSFET represents an increase in power of almost three and a half times for a price factor of two. When compared with lower powered devices around 10A the power increase to price increase using the same comparison is a twenty one times increase in power for four times increase in cost.

The secondary effect of using larger drive components is that they are less stressed and therefore often more reliable.

Stepper motors

Using high voltage drive techniques may mean that a smaller stepper motor can be used in a project.

Microprocessors

The use of microprocessors often means a greatly reduced component count. This impacts on ancillary parts such as PCBs and enclosures. Secondary sources of components Unfortunately in many respects we are living in an age of disposable goods. Technology moves on in many areas but some items change very little.

Motors have fundamentally changed little for many years. Redundant computer equipment such as printers and the older types of disc drives are often a source of small stepper motors. It is sometimes less expensive to buy a new item and strip for parts than to build from the

A range of power MOSFETs and power transistors.

component parts. An example of this is the pulse width servo. For an application requiring a powerful servo drive an inexpensive standard servo was used. This was a 'bargain' type from a local model shop. The retail price was about £5. It was this price because the motor and drive were bronze bushed not ball bearing drive. This did not matter because the motor and drive were discarded.

The motor drive signals were used to feed high power drivers and the internal potentiometer was replaced by an external potentiometer and connected back to the servo unit. If the unit should fail it is not a great problem. This is because these units are plug and play it does not matter if the control IC is superseded the repair of the unit is simply another 'cheap' servo unit.

Overcoming obsolete component failures

All components have a finite production life. This causes problems with older circuits or repairs.

One of the examples I was going to use in this book for the simple driving of stepper motors was the SAA1027. This is a device that has been about since the 1980s and was in most retail catalogues until recently. During the writing of this the manufacturers asked for 'last orders' on the device and it appears to be no longer available through normal retail sources. Even if available it would not be prudent to construct a design using a component that has a limited life. The SAA1027 does not seem to have a direct specification replacement and certainly not a pin compatible replacement.

There are a number of options if a repair to or construction of another unit involving obsolete components is necessary.

Scrap

If the electronics unit is simple it may be best to cut your losses and scrap the controller part of the electronics. Items such as power supplies, motors, driver MOSFETs, driver transistors and control hardware are likely to be unaffected. If the unit had been well designed initially most of these items will have plugs and sockets to connect to the control PCB and any high current units will probably be directly connected to the power supply.

Piggyback

On complex units ICs can be replaced by

building a PCB section that plugs into the original IC socket or a socket used to replace the faulty IC. The PCB will contain components and circuitry that simulates the replaced item.

Don't go there in the first place
For the 'home constructor' this is probably the simplest option. It means careful planning of the components used.

Instead of using one large PCB, split the PCB into separate 'task' sections.

Where possible, use as many type components as possible. These are components that do not do specific tasks but are for want of a better phrase are 'jack of all trades'. Consider the 741 OP amp this one of the early commercially successful and affordable ICs. It has appeared in innumerable circuits and designs. If the 741 stopped being made tomorrow it would matter very little because there are a number of later devices that were based on the same format and are pin compatible.

The ULN2803 is a common device used for driving outputs. This could feasibly, but not as neatly, be replaced by individual resistors, small drivers and diodes if the device ever went out of production.

Microprocessor devices are unusual in that many of them are generic type base devices. It is the programming that gives these devices uniqueness. Mask programmed devices are produced in large quantities by the manufacturer when the device is made but burnt in and flash programmed devices can be produced for a specific task in small quantities – as low as one item. These types of devices regularly evolve but usually remain fully compatible. The manufacturers of the generic type base devices are unlikely to cease production of these devices and often there are a number of other sources or pin compatible devices available.

Low stress design

This is about thinking about ways of protecting the 'higher level' components. Many microprocessors are capable of sourcing and/or sinking currents in the order of 25mA. This is the

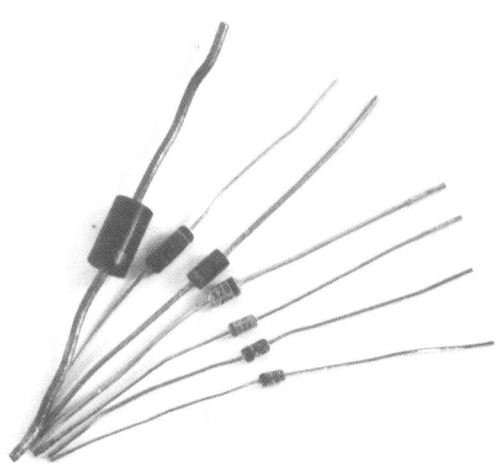

A range of signal diodes.

capability to drive directly an indicator LED or a reed relay and would probably be used for this in a commercial product. Any failure or short on the driven device is likely to result in the 'blowing' of the microprocessor. Using a single transistor driver or an integrated driver such as the ULN2803 could protect against this failure. Although the short may result in the failure of the ULN 2803 this is of less consequence than failure of the microprocessor. The few extra components used are generally of minor concern in designs not intended for commercial manufacture where price is often a major criterion.

Propagation delays

When any electronic device switches there is always a delay between the input and the output effect. This is known as the propagation delay. The value of this depends on the device type and also the device family. The value is also affected by other factors such as the supply voltage and the load capacitance. Typically for high speed CMOS logic this is in the order of 10nS and for 4000 CMOS is in the order of 90nS at 5V supplies.

Propagation delays are usually regarded as a nuisance limiting the maximum speed that a circuit can operate. There is also a plus side in that propagation delays can be used to ensure that critical timing operations occur in the correct order. In circuits where multiple events are triggered from the same input errors can occur if the transition time of one of the input events is quicker than another. In this case a gate or string of gates can be used to provide a delay to subsequent events to ensure that events that need to occur first are at a position for the subsequent event to occur.

Propagation delays can be produced from many different types of gates and combination of gates.

Fig.274 shows a series of buffers used to produce a series of delays. When the input is taken positive each subsequent stage will go positive after the propagation delay. The total delay is the sum of each individual delay.

Fig.275 shows a series of inverting buffers used to produce a series of delays. When the input is taken positive each subsequent stage will go the inverse of the input to that stage after the propagation delay. The total delay is the sum of each individual delay.

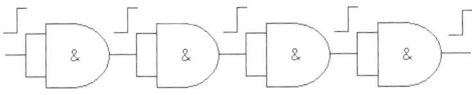

Fig.276 shows a series of two input and gates wired as a buffer to produce a series of delays. When the input is taken positive each subsequent stage will go positive after the propagation delay. The total delay is the sum of each individual delay.

Basic gates such as and, nand, or, and nor and the Schmitt trigger versions of these where available can be connected to produce inverting or non-inverting stages depending on their type. Two inverting gates in series produce an overall non-inverting effect.

Small capacitors between the stage outputs and ground can be used to increase the propagation delay.

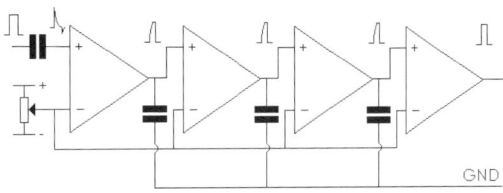

Fig.277 shows the method of using an OP amp/ comparator as a Schmitt trigger for producing propagation delays. The input capacitor is used as a pulse shortener and the intermediate stage capacitors are used to increase the propagation delay. The shape of the intermediate pulses will not be square due to the capacitor. If a sharp rise time is important the capacitor on that stage can be left off or the output can be put through another Schmitt trigger stage.

Practical uses of propagation delays

Beside the timing modification uses it is possible to use the delay to produce a 'travelling pulse' to trigger a series of events.

If an input capacitor is used as a pulse shortener the initial pulse produced will travel along the 'delay line'. The output can be taken at various stages to trigger events. Some experimentation will be necessary to find the exact delay times suitable for the application particularly if one event must end before the next commences.

Practical electronic building blocks
The driver circuits in this book can all be built to a limited number of patterns making the

production of finished units extremely simple. The standard pattern is a piece of aluminium plate with up to four drivers mounted. The majority of drivers used are TO220 mounts. Four drivers on a heatsink allow the unit to be a full 'H' bridge driver. One of these will be necessary for a motor 'H' bridge or unipolar driver and two will be necessary for a bipolar stepper motor drive. Freewheeling diodes can also be fitted to the unit.

The same unit with a minor change in wiring is a four output unipolar driver. One of these will be necessary to drive a unipolar stepper motor.

The same unit with other minor changes in wiring is a double output MOSFET driver with freewheeling MOSFET and brake MOSFET. If the second MOSFET of the double output driver or the brake is not required these can be left off.

All the above units can also have the gate to ground resistors built in. This means that the only electronic connection to the unit is the on switching.

If high current or high heat dissipation devices are used then the size of the standard heat sink unit can be scaled up by a corresponding amount or one of the other techniques discussed in the book can be used.

Making timing strips and discs

Timing strips and discs can be made in a number of materials depending on the counting sensor. Simple single or low number slot strips and disc suitable for use with simple optical or magnetic devices can be made by milling or cutting slots in brass sheet or strip.

When the number of timing marks increases then simple mechanical cutting is no longer suitable. Most timing strips or discs are produced photographically or use photographic processes as an intermediate stage for etching brass or stainless steel sheet.

Practical considerations

Most multi 'slot' timing strips or discs will use optical methods of reading. It is usual to use a series of light and dark areas of equal width or slot/ no slot if etched.

The slot in the sensor must be narrower than the divisions of the timing strip or disc. This is to prevent light passing through a number of slots and giving false readings. Also the sensor assembly should be in a dark or sealed environment with the nearby areas being coloured matt black to prevent extraneous reflections giving false readings.

These considerations become more important the higher the density of slots becomes.

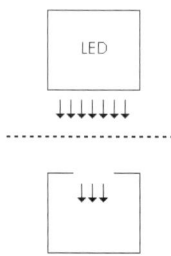

Fig.278 shows the effect of using a sensor with a wide aperture. Light from a number of timing slots is being sensed.

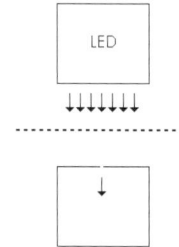

Fig.279 shows the effect of using a sensor with an aperture narrower than the timing slot. Light from only one timing slot is being sensed.

Double clocking

With timing strips and disks there is normally only one output pulse per slot/ no slot pair. If the slot and no slots are of equal dimensions and the count is only in one direction double clocking can be used. This is where the electronics produce an output on both the fall of clock and rise of clock i.e. on the transition from light to dark and from dark to light.

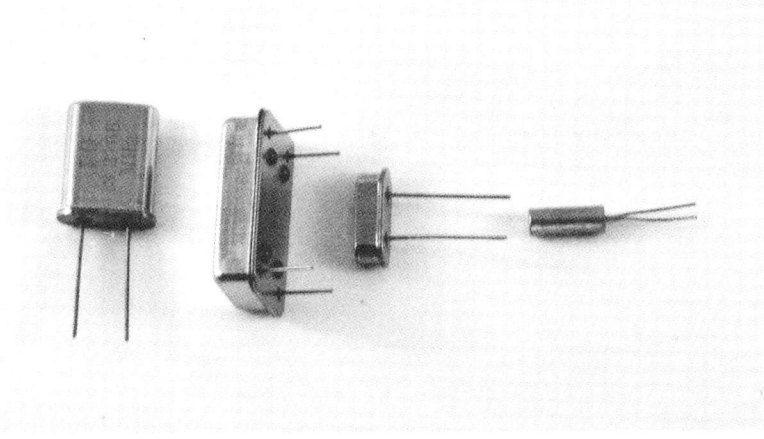

A range of crystal oscillators.

The advantage of double clocking is that it doubles the effective count from a timing strip or disc and hence doubles the resolution. This means that a disk can be produced with half the number of slots. As an example consider an incremental sensor disc to produce an output for a rotary device with a required resolution of 1°. With single clocking this would require 360 slots. With double clocking this same disc would produce o resolution of 0.5° or the number of slots in the disc could be reduced to 180.

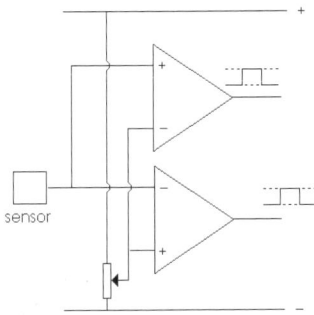

Fig.280 shows a method of obtaining double-clocked outputs. The outputs shown would then be fed through some form of OR gate to give a single output. Depending on the speed and type of optical sensor used it may be necessary to use an offset on the comparators or two potentiometers to provide individual compare points. Without these precautions the output could be a continuous output instead of discrete steps because one comparator could switch on before the other switched off. The offset techniques are described in the chapter on servo drives.

Absolute and relative encoders

Absolute encoders produce at switch on an output with the correct reading. Relative encoders need to move through a reset position at power up or they can be used to read relative to a chosen start position. They work by adding or subtracting the count from the arbitrary or set start position.

Relative encoders are the simplest and easiest to produce. For the example quoted of the 360° disc with double clocking only 180 slots are needed on one track. If absolute positioning were

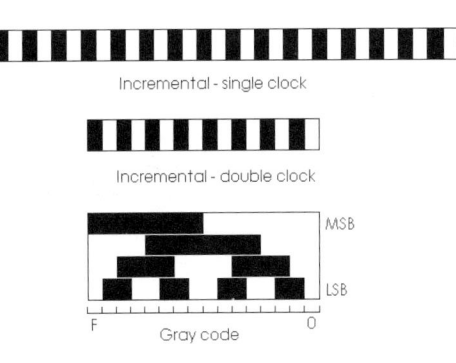

Fig.281

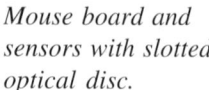

Mouse board and sensors with slotted optical disc.

required a second track slot would be needed to provide a home position and this home position would have to be passed through to start the counting.

With true absolute encoding double clocking is not appropriate because absolute encoding uses definite levels not clock transition. A Gray code absolute encoder for 360° would require 9 bits of output and therefore 9 tracks of holes. This may appear difficult if an attempt is being made to produce a timing disc but in common with most binary counting the number of bits in each track average 50% of the full track count and the gaps and lines become in effect larger moving from LSB to MSB. With multiple track encoders a sensor will be needed for each track or bit. Decoding of the output is easy using the techniques explained in the section on binary and Gray code conversion.

Fig.281 shows a single clock incremental timing strip and a double clock incremental timing strip compared to a Gray code timing strip. For simplicity the count is limited to 0 to 15 i.e. 0H to FH. Direction information for the incremental strip would be by using offset sensors or if the strip was being used in conjunction with a motor drive the up/ down information could be obtained from the motor logic.

Practical up/ down counting

Commercial disc and sensor units are available giving hundreds and sometimes thousands of counts per revolution but even in simpe form they tend to be relatively expensive.

As mentioned previously in the section on sensors, up/ down counting presents a number of problems.

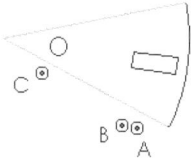

Fig.282 diagram shows a section of a circular timing disc. Only one hole is shown. A and B are offset sensors. If an output from sensor A is followed by an output from sensor B the disc is moving clockwise. If an output from sensor B is followed by an output from sensor A the disc is moving counter clockwise. Sensor C is the home position sensor. This is used to keep the count in synchronisation. There is only one hole in the disc to represent home position. Problems occur with this simple set up if the disc stops exactly

Mouse board and sensors with transmissive optical disc.

over the sensors and then moves back in the direction it was travelling from. The effect will be extra counts occurring without the sensor disc actually moving to the next disc position and therefore after a short time the disc count will be out of synchronisation with the physical position. This is common in simple systems where the timing disc is used with a mechanical system with 'backlash'. Simple systems used in the drive of a machine tool can result in cumulative errors and a pile of scrap.

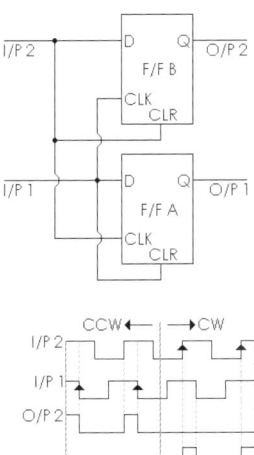

Fig.283 shows a simple count decoder. The circuit is not foolproof and accuracy depends to some extent on the sensor spacing and timing disc slot size. Extra logic gating can be added to make the circuit more foolproof. The first step is to use input signal conditioning in the form of Schmitt triggers or OP amp voltage comparators to produce clean switching transitions.

The circuit use 2, D type flip-flops. With a D type flip-flop the data on the data input D is transferred to the output Q when a pulse appears on the clock input CLK, and the clear input CLR is not activated. A suitable IC for this circuit is the one of the low power versions of the dual D flip-flop 7474 such as the 74HC74. With this IC the clock inputs are active high rising edge i.e. function with a true or '1' input rising. Clear is achieved by a low on the clear input.

The directions CCW and CW are notional; the actual direction will depend on the unit design. Consider the clockwise direction.

With inputs I/P 1 and I/P 2 both low, the clear functions are active and there is no output on O/P 1 and O/P 2.

When I/P 1 goes high, F/F B is clocked but there is no D input therefore both outputs remain low. When I/P 2 goes high, F/F A is clocked and there is D input on F/F A from input I/P 2; therefore output O/P 1 goes high.

When I/P 1 goes low, O/P 1 is cleared.

Suitable sensors

There are a very limited number of units available

with two offset sensors mounted in one package. One commercially available unit is the HLC2701. It is possible to produce 'home-made' sensor units for special applications using optical fibre feeding a standard optical sensor. 1mm optical fibre can be easily squeezed into an oblong shape and mounted with an offset using epoxy resin. The light pick up surfaces are then polished to allow the maximum light into the guide.

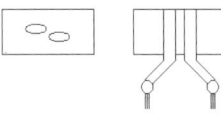

Fig.284 shows a possible structure of an optical sensor using light guide and standard opto sensors. The most common application of up/down counting is probably the computer mouse. These counters work extremely well but loss of count in this application is of little importance because the 'system' is a hand/ eye/ brain synchronisation where errors are corrected automatically by the person 'driving' the mouse. The mouse is produced in such vast quantities that the price is low.

An option is to use the sensors from the mouse. Each mouse contains two sensor/ led units that can be used with the existing mouse logic. Because these circuits vary it is a matter of experiment to produce a working unit.

Another option for up/ down counting is to use just the sensor/ led units with the count logic described earlier.

CHAPTER TWENTY ONE

Etching Processes

Most of the following section applies equally well to the production of PCBs and etched metal timing discs.

Photographic method

Black and white film can be used to produce timing strips and disks by the contact method. This requires little in the way equipment. Slow films can be obtained that do not even require a darkroom but do need long exposure times. A master will be required, but once this is produced any number of copies can be made. Copyright laws will apply if the master is produced from a commercial design.

The film can be used directly as the timing strip or disc but care must be taken with positioning the film relative to the sensor/ emitter mounting to prevent scratching as the film and sensor/ emitter move relative to each other.

PC and PC printer

Many people have access to a PC. These often have simple bitmap drawing programs included in the standard software. The main problem with BMP type programs is the coarseness of pixel blocks resulting in angled lines and curves that are coarse and lumpy. The other problem is that of accurate scaling of the design to a specific size.

There are a number of inexpensive vector drawing programs available that overcome these problems.

Master images can be produced directly from the printer particularly a laser printer. The image from a laser printer can be printed onto many kinds of thin white paper or onto OHP film.

With the common and inexpensive inkjet printer the ink is very fast drying and often leads to areas of black appearing in the form of a series of stripes. This is of no use with most small area optical sensors.

If a print is made on OHP film with an ink jet printer the areas of black will appear streaked but if the printed side is held *carefully* over the steam from a kettle, the ink returns to a semi liquid form and surface tension pulls the lines and black areas into consistent black areas with clean edges.

Metal etching

Metal timing strip or discs are usually made of metal thick enough to be 'free standing' but metal foils are available that can fixed to a clear backing using a clear adhesive.

The process of metal etching is similar to making PCBs. Depending on the type of metal being used the same processes can be applied.

Undercutting during etching

The chemical etching of metal sheet does not produce vertical cuts into the metal. An undercut occurs that is usually quoted as a percentage of the metal thickness. Many factors can have an effect on the final shape. These include temperature, agitation and width of slot compared to metal thickness. The last two factors can lead to localised chemical depletion slowing down the rate of etching particularly where undercuts have started.

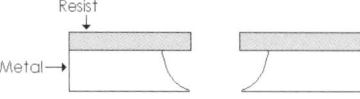

Fig.285 shows a typical single undercut due to the etching chemicals acting on the metal exposed as the etching proceeds on the thickness of the metal during a single side etch. The resist on the back of the metal is not shown. With a PCB the backing would be the board the copper foil was glued to.

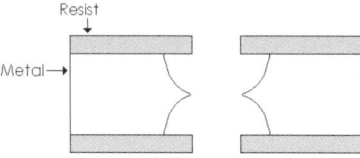

Fig.286 shows a typical double undercut due to the etching chemicals acting on the metal exposed as the etching proceeds on the thickness of the metal.

Fig.287 shows the area where the undercut can vary due to localised chemical depletion.

Photo chemicals and plates

Chemicals and plates for both PCBs and metal etching are available commercially. Plates and metal sheet is available ready coated. Problems can occur if it is required to coat the surface of a metal plate manually. Spray photographic resists are readily available but producing an even coat is often found difficult. In commercial laboratories quality testing plate and chemicals, manual coating is an everyday occurrence. The process used is to put the plate on a horizontal centrifuge and apply liquid resist slowly.

Etching processes

As with all chemicals safety precautions should be observed and is important that safety notices on chemicals should be read, understood and strictly adhered to.

For occasional use Ferric Chloride is probably the most commonly used chemical etchant. The initial chemical reaction for Copper is
$FeCl_3 + CuCl = FeCl_2 + CuCl$
As the etching proceeds the CuCl produced also acts as an etchant and two reactions take place.
$FeCl_3 + CuCl = FeCl_2 + CuCl$
$CuCl_2 + Cu = 2CuCl$
Both these reactions are oxidation reactions and therefore the reaction is quicker in the presence of oxygen.

Bubbling air through the reactants helps to carry away depleted chemicals and provide oxygen. This speeds up the reaction greatly as does raising the temperature. But both actions can result in the production of unpleasant fumes. Therefore the reaction should not take place in a confined space.

There is a problem with staining and disposing of spent solutions.

Ferric chloride will also etch brass and stainless steel with a similar oxidation process.

There are a number of other chemicals that perform etching in a similar way to ferric chloride. These are sodium persulphate, which is readily available and ammonium persulphate, which is less readily available. Both suffer from disposal problems.

Another widely used process for the etching of copper is to use Cupric Chloride in the presence of Hydrochloric acid. This is used because the chemical process is reversible, which means no waste disposal problems.

A typical recipe used is
Cupric Chloride as a solid $CuCl_2$
200gms

Hydrochloric acid 37.5%   HCl
100gms
Water H_2O
1000ml
The chemical reaction is
$CuCl_2 + Cu = Cu_2Cl_2$
The colour of the solution changes from bright green to brown.
With the addition of oxygen a secondary reaction occurs.
$2Cu_2Cl_2 + 4HCl + O_2 = 4CuCl_2 + 2H_2O$
The method is to blow air through the chemicals during the etching process and continue blowing air through after the etching is complete and the plate removed until the colour returns to the original bright green. There is less of a problem getting rid of spent solutions because the solution is much longer lasting because the chemical reaction is reversible. Solutions only need to be disposed of if contamination occurs.

Electro chemical etching

This is a process that can only practically be used for etching because the process ceases in areas that are not in continuous contact with the power supply connection. If the process were used for PCB production, 'island' areas of unetched copper would be likely as areas were etched away that were the connection to other areas. With a PCB the area of remaining copper is probably only a small percentage of the original copper area. On a metal etched sheet the opposite occurs, the area of metal etched away is usually only a small percentage of the original metal area.
The process uses easily available chemicals i.e. water and salt. The process is in effect electro plating in reverse. Unprotected areas of metal connected to the positive of the power supply are deposited on the metal electrode connected to the negative of the supply. This results in the unprotected areas being 'eaten' away. The process will work with a large range of metals. The system is much safer than those involving chemicals with a disposal problem. The salt solution acts as a carrier for the current and remains unchanged at the end of the process. Fig.288 shows a set up for single and for double side etching. The plate to be etched is connected to the positive rail and the plates that act as receivers for the plated metal are connected to the negative. The voltage is not critical and a voltage of between 6V and 12V is adequate. The power supply needs to be able to supply a current of about 5amps for the full period of etching. A car battery is ideal for this. With plating excessive currents can produce a discoloured surface. With etching the appearance of the receiver plate is not important therefore fairly large currents can be used leading to a reduction in the etching times. There is always a balance and the current must not be so large that it causes lifting of the resist coat at the edges of the sections being etched away. With electroplating it is normally a more 'noble' metal that is being plated on a less 'noble' metal e.g. copper or tin on steel and silver on copper or brass. Consider the plating of silver on copper, initially there are two distinct metals in the reaction i.e. silver and copper. As the process continues, and a layer of silver is built up the reaction is actually silver being plated on to silver.
With etching a similar piece of metal for the receiver to that being etched is used e.g. stainless steel with stainless steel and brass with brass.

Using master images

When using a master image to produce secondary images it is important that the printed image is in contact with the photo resist coating directly. If the image is separated by the thickness of the film or paper being used for the master then the edges of the image produced may not be sharp or may be distorted due to the angle of the light used for developing or extraneous and reflected light. It may be necessary to print the image reversed to obtain the correct alignment. Printing a symbol or marker on the master e.g. © and a name will help to prevent orientation 'accidents' when timing strips and discs are being exposed.

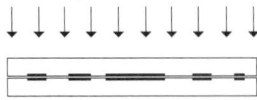

Fig.289 shows the correct orientation of the master image and the required image.

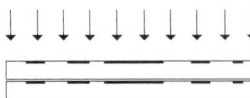

Fig.290 shows the incorrect orientation of the master image and the required image.

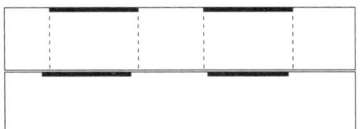

Fig.291 shows exaggerated on a small part of the exposed image the distortion that can occur by using the incorrect orientation of the master image and the required image. The actual effect will depend on the type and orientation of the light source. The effects are 'shadowing' that can produce shifts in the image edge. This can result in oversize and undersize 'tracks' relative to the master. The effect may also result in a total 'track' shift. The effect is magnified if thick film is used because of the greater separation and by light travelling obliquely within the film.

This effect applies equally to PCB production and to timing strip and disc production. The effect is more noticeable with fine line timing strip and disc production because of the effect may result in a non-linear count.

The effect is less noticeable with PCBs unless fine tracks with narrow separation are being used.

Aligning double sided images

The aligning of two images on opposite sides of a piece of PCB board is difficult with the normal amateur way of making PCBs.

With amateur PCBs the holes are drilled after etching. Professionally the PCB is drilled as the first main operation in the production of the PCB. The two full layers of copper are kept intact late into the production process to be available for processes such as through plating of the holes and for the electroplating of solder on to the tracks. Alignment holes are often used particularly in short run or prototype production. It becomes immediately obvious if there is an alignment problem because of the large number of holes needing to match up.

For 'one offs' it is easy to use a series of strategically placed hole to help with the alignment.

Metal sheets

These present a more difficult problem than PCBs because the metal sheet is often thin and flexible.

Alignment holes can be used but a common method is to align the two image sheets face to face and slide the metal sheet between.

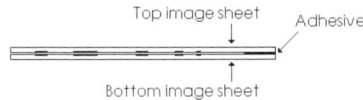

Fig.292 shows two image sheets attached together with an adhesive layer such as double-sided adhesive tape.

The image can be accurately aligned visually before applying the adhesive layer.

This is a simple method but can lead to a misalignment because of the method of fixing if relatively thick metal is being used.

Fig.293 shows the effect on the alignment when using a metal plate that is thicker than the adhesive layer.

Fig.294 shows the effect on the alignment when using a packing piece that is the same thickness as the metal plate.

PCB making

It is not intended that this book will go into any depth on the making of PCBs. There are many excellent books on the market covering production on both a professional level. Master images can be made up in a number of ways from drawings, photocopies, 'rub down' stencils and tapes. Simple PCBs can be made by drawing directly onto the board with etch resist pens although the results are not always predictable.

The best way is probably to use a PCB drawing package. These have dropped enormously in price and improved in quality over the last few years. There are packages intended for the amateur that do not have features such as Gerber outputs.

CHAPTER TWENTY TWO

Using Stripboard For Prototypes

Strip board is the favoured material of many people for building 'one off' or prototype circuits. It is available in SPBP of fibreglass board. The fibreglass is a little more expensive but is stronger and maintains better insulation between tracks. There are many layouts for the stripboard copper tracks but the most common is a series of parallel conductor strips with holes at 0.1in. centres. Designs can look neat or very messy depending on the approach. The messy layouts can also have inherent problems such as cross talk between adjoining tracks.

Method of using

This is just one method of using stripboard, there are many more.
Work from the non-tracked side of the board. Do not try to make prototype boards too small this can lead to layout difficulties and testing and modification problems.
Establish power rails. Depending on the circuit this may be a positive on one edge of the board and a negative on the other edge. For more complex circuits it is better to have both a positive and a negative rail at each board edge. For multiple IC circuits it is better to have a pair of rails between each row of ICs.
Mark the rails red and black using an indelible marker on both sides of the board. These measures cut down on the amount of wiring and the marking cuts down on errors.
Make layouts on paper or from plug in prototype boards if these are used for circuit testing.

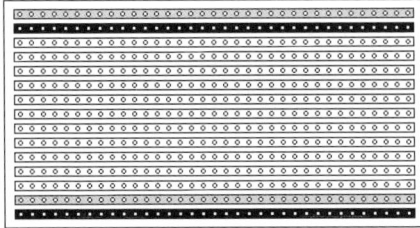

Fig.295 shows the stripboard from the copper side with power rails marked.
Next decide on the layout of the IC sockets. It is usually better to use sockets for ICs because repair is easier and if the design does not work as predicted it is easy to salvage the ICs and try again. Cut the tracks between the pin rows and between adjacent ICs before soldering the sockets in place.

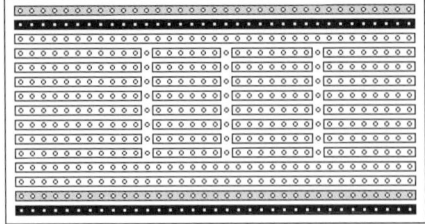

Fig.296 shows the stripboard from the non-copper side with the tracks cut and the power rails marked.

Connect the power pins of the ICs to their relevant rails and any other pins that need to be tied to the rails. Use wires of different colours to help trace the circuit through if there are problems.

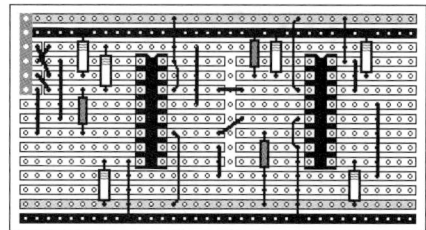

Fig.298

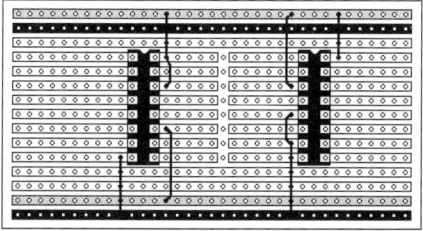

Fig.297 shows the layout from the non-copper side with sockets in place and first wiring.
Then lay in small components such as resistors, capacitors and connectors.
Fig.298 shows the layout from the non-copper side with sockets, resistors, capacitors and connectors in place. Where possible use any spare tracks as interconnects instead of wires. It is now easy to add any extra wires. Also strap any unused gates to a rail as per manufacturers recommendations. Finally add any large components such as electrolytic capacitors that block access to the board.
Neat wiring makes testing, fault finding and any future modification easier.

CHAPTER TWENTY THREE

Pin Outs & Specifications

It should be noted that variations occur with specifications of apparently similar devices. This may be due to different manufacturers or may be because of different packaging types. This variation is often denoted by a suffix. The table information is a 'typical' specification. The tables only cover a small percentage of available types but lists types that should be easily available.

High power MOSFETS
The following table gives a list of power MOSFETs shown in order of ascending current capability.

High power MOSFETS						
Device	Channel	Voltage	Current	Watts	Ohms	Case
IRF520	N	100	10	70	0.2	TO220
IRF640	N	200	18	125	0.18	TO220
IRF540	N	100	30	100	0.052	TO220
HUF75939P3	N	100	53	200	0.02	TO220
HUF75337P3	N	55	62	115	0.014	TO220
IRF4905	P	55	64	150	0.02	TO220
HUF76145P3	N	30	75	325	0.0045	TO220
IRF2907	N	75	209	470	0.0045	TO247

Low power MOSFETS
The following table gives a list of power MOSFETs shown in order of ascending current capability.

Low power MOSFETS						
Device	Channel	Voltage	Current	Watts	Ohms	Case
VN2410L	N	240	0.18	1	10	TO92
2N7000	N	60	0.2	0.35	5	TO92
VN10LP	N	60	0.3	0.62	5	ELine
VN2222LL	N	60	0.5	0.4	7.5	TO92
IRFD210	N	200	0.6	1	1.5	HD1
IRFD220	N	200	0.8	1	0.8	HD1
3N163	P	40	0.03	0.37	0.25	TO72
ZVP3306A	P	60	0.16	0.62	14	ELine
ZVP4424A	P	240	0.2	0.75	9	TO92
IRFD9110	P	100	0.7	1.3	1.2	TO250

Power transistors

The following table gives a list of power transistors shown in order of ascending current capability.

Power transistors						
Device	Channel	Voltage	Current	Watts	Gain	Case
MJE340	NPN	300	0.5	20	30-240	TO126
TIP31C	NPN	100	3	40	10-50	TO220
TIP41C	NPN	100	6	65	15-75	TO220
MJE15028	NPN	120	8	50	40+	TO220
MJE3055T	NPN	60	10	75	20-70	TO220
TIP35C	NPN	100	25	125	10-50	SOT93
MJE350	PNP	300	0.5	20	30-240	TO126
TIP32C	PNP	100	3	40	10-50	TO220
TIP42C	PNP	100	6	65	15-75	TO220
MJE15029	PNP	120	8	50	40+	TO220
MJE2955T	PNP	60	10	75	20-70	TO220
TIP36C	PNP	100	25	125	10-50	SOT93

Darlington power transistors

The following table gives a list or Darlington drivers shown in order of ascending current capability.

Darlington power transistors						
Device	Channel	Voltage	Current	Watts	Gain	Case
BD679A	NPN	80	4	40	750+	TO126
TIP121	NPN	80	5	65	1000+	TO220
TIP132	NPN	100	8	70	1000-1500	TO220
BDX33C	NPN	100	10	70	750+	TO220
BDV65	NPN	60	12	125	1000+	TO218
BDW42	NPN	100	15	85	1000+	TO220
MJ11016	NPN	120	30	200	1000+	TO3
BD678A	PNP	60	4	40	750+	TO126
TIP135	PNP	60	8	70	1000-1500	TO220
BDX34C	PNP	100	10	70	750+	TO220
BDV64	PNP	60	12	125	1000+	SOT93
BDW46	PNP	80	15	85	1000+	TO220

Low current rectifier diodes

The accompanying table gives a list of rectifier diodes shown in order of ascending current capability. These diodes are relatively low frequency devices suitable in applications such as power supplies and low frequency blocking.

Device	PIV	Current amps	VF drop volts
1N4001	50	1	1.1
1N4002	100	1	1.1
1N4007	1000	1	1.1
1N5400	100	3	1.1
1N5402	200	3	1.1
1N5408	1000	3	1.1

Schottky barrier rectifier diodes

The following table gives a list of rectifier diodes shown in order of ascending current capability. These diodes are relatively low frequency devices suitable in applications such as power supplies and low frequency blocking.

Device	PIV	Current amps	VF drop volts
1N4001	50	1	1.1
1N4002	100	1	1.1
1N4007	1000	1	1.1
1N5400	100	3	1.1
1N5402	200	3	1.1
1N5408	1000	3	1.1

Fast rectifier diodes

The following table gives a list of fast diodes shown in order of ascending current capability. These diodes are fast devices suitable in applications such as freewheeling diodes for inductive loads and high frequency blocking and rectification.

Device	Max V volts	Current amps	VF drop volts
BAT41	100	0.1	0.45
1N5819	40	1	0.55
1N5820	20	3	0.85
1N5821	30	3	0.38
1N5822	40	3	0.52
PBYR10100	100	10	0.7

Ultra fast rectifier diodes

The following table gives a list of fast diodes shown in order of ascending current capability. These diodes are fast devices suitable in applications such as freewheeling diodes for inductive loads and high frequency blocking and rectification.

Device	Max V volts	Current amps	VF drop volts
BYV26C	600	0.65	2.5
1N4937	600	1	1.1
BY396P	100	3	1.25
1N3890R	1000	12	1.4
MUR1540	400	15	1.25
BYW96E	1000	30	1.5
70HFLR60S02	600	70	1.85

Signal diodes

The following table gives a list of signal diodes. These diodes are suitable in applications such as producing voltage offsets, logic gating and signal routing. The case styles and pin connections are shown after the table.

Device	Max V volts	Current amps	VF drop volts
BYW29E-200	200	7.3	1.3
MUR860	600	8	1.5
BYV79E-200	200	12	1.3
MUR3060PT	600	15	1.5
BYV32200	200	20	1.2
BYT200PIV400	400	100	1.6

Regulators

There are numerous types and specifications of regulators. These are often in a TO92 style cases for the low current devices and TO220 style case for the one and two amp devices. A few of the larger current devices may be in packages such as the TO3. The majority of devices use a standard connection layout for a specific case style.

Device	PIV	Current amps	VF drop volts
1N914	75	0.075	1
BAT46	100	0.15	0.45
BAT86	50	0.2	0.38
1N4148	75	0.2	1
BAW62	75	0.25	1
BAT47	20	0.35	0.8
BAT49	80	0.5	0.42

Voltage step up devices

There are a number of ICs that can produce stepped up voltages. These are useful for driving solenoids or MOSFETs from lower voltages. Some devices such as the LM2577 are in effect switch mode drivers. They can provide a regulated output at up to 100V with current capabilities of 3A. They use a number of external components to achieve the voltage step up. Devices such as the 7660 use a minimum of external components to give either a complimentary voltage up to 12V. I.e. an input of +12V gives an output of −12V. They can also be used in a voltage doubling configuration where a +12V input will give a +22.8V output. The output current is limited but should be capable of

Pin Outs & Specifications

A prototype strip board layout for a PC I/O board.

driving a number of MOSFET driver stages. There is also a number of DC isolating charge pump convertors that can give voltages typically of plus and minus 12 to 15V for an input of about 5V. These types of devices give currents in the order of 50mA. They are relatively expensive and are usually used in battery circuits or for localised power supplies in processor circuits where the additional expense is outweighed by the advantages over fitting more complex power supplies and wiring.

Simple inverters can be made using 555s and a ferrite ring wound as a step up transformer.

Because most of these devices are oscillators, the LM2577 runs at 52KHz, induced noise from the oscillator can be a problem without efficient filtering. In most units not intended for manufacture and for 'one offs' it is probably easier to add extra windings to a transformer or add a secondary transformer to provide the low current supplies. With battery circuits there is often little choice but to use voltage converters. These devices are explained in the power supply sections.

ULN2803

This device is included because of its versatility and ease of use. The device has eight outputs that can individually supply up to 500mA and can be connected in parallel to provide up to 4A. It is a self contained unit in that it contains base drive resistors, base to ground resistors, base to ground protection diodes and freewheeling diodes for inductive loads in a single package. The output voltage is equivalent to the supply voltage and can be between 3 and 50V. The only drawback with this device is that it can only sink current not source it.

CHAPTER TWENTY FOUR

Information Sources

The worldwide web can be a good source of information although the validity of information from a small number of sources is not always proven or accurate.

Search engines are becoming more discriminatory even with large companies. Failure of one search engine to find a piece of information does not mean that the required information is not on the web.

Putting a generic IC number into some search engines results in a blank or sometimes a diverse unrelated collection of responses.

The URL of a company may be not obvious from the name.

The following is a list of manufacturers' websites with URLs. These are not recommendations but simply information and data sources relevant to this book.

Most of the larger manufacturers only supply in large quantities or to special order.

Web sites may also list retail outlets or in the case of multinationals, list their local offices providing information on local suppliers of their product.

Manufacturers who make to special order are often willing to put you in contact with their customers who hold stock and are willing to supply in small quantities.

Generic components are available from most high street electronic suppliers.

allegromicro.com
Hall effect and current sensors and other IC ranges
bluemole.co.uk
Control microprocessors, self assembly kits and plans
bourns.com
Comprehensive product range including switches and rotary encoders
cherrycorp.com
Switches, keyboards and magnetic sensors
emd.co.uk
Fractional horsepower DC permanent magnet electric motors and worm gearboxes.
ex.ac.uk
This site provides a summary of most of the units of measurement to be found in current and historical use.
fairchildsemi.com
Large range of semiconductor products
hengstler.com
Encoders and associated products
intersil.com
Large range of semiconductor products
kingbright.com
LEDs and optical devices
lumileds.com
LEDs and optical devices
mechetronics.co.uk
Solenoids and solenoid valves
national.com
Large range of semiconductor products
nichia.com
LEDs and optical devices
osram-os.com
LEDs and optical devices
relays-r-us.co.uk
Comprehensive source of relay data
saia-burgess.com
Switches, motors and solenoids
ti.com
Large range of semiconductor products
tycoelectronics.com
Comprehensive source of relay data